# Advances in Intelligent Systems and Computing

Volume 1423

The series "Advances in Intelligent Systems and Computing" contains publications on theory, applications, and design methods of Intelligent Systems and Intelligent Computing. Virtually all disciplines such as engineering, natural sciences, computer and information science, ICT, economics, business, e-commerce, environment, healthcare, life science are covered. The list of topics spans all the areas of modern intelligent systems and computing such as: computational intelligence, soft computing including neural networks, fuzzy systems, evolutionary computing and the fusion of these paradigms, social intelligence, ambient intelligence, computational neuroscience, artificial life, virtual worlds and society, cognitive science and systems, Perception and Vision, DNA and immune based systems, self-organizing and adaptive systems, e-Learning and teaching, human-centered and human-centric computing, recommender systems, intelligent control, robotics and mechatronics including human-machine teaming, knowledge-based paradigms, learning paradigms, machine ethics, intelligent data analysis, knowledge management, intelligent agents, intelligent decision making and support, intelligent network security, trust management, interactive entertainment, Web intelligence and multimedia.

The publications within "Advances in Intelligent Systems and Computing" are primarily proceedings of important conferences, symposia and congresses. They cover significant recent developments in the field, both of a foundational and applicable character. An important characteristic feature of the series is the short publication time and world-wide distribution. This permits a rapid and broad dissemination of research results.

Indexed by DBLP, INSPEC, WTI Frankfurt eG, zbMATH, Japanese Science and Technology Agency (JST).

All books published in the series are submitted for consideration in Web of Science.

For proposals from Asia please contact Aninda Bose (aninda.bose@springer.com).

More information about this series at https://link.springer.com/bookseries/11156

Yasufumi Takama · Naohiro Matsumura ·
Katsutoshi Yada · Mitsunori Matsushita ·
Daisuke Katagami · Akinori Abe ·
Hisashi Kashima · Toshihiro Hiraoka ·
Takahiro Uchiya · Rafal Rzepka
Editors

# Advances in Artificial Intelligence

## Selected Papers from the Annual Conference of Japanese Society of Artificial Intelligence (JSAI 2021)

Springer

*Editors*
Yasufumi Takama
Graduate School of System Design
Tokyo Metropolitan University
Hino, Tokyo, Japan

Naohiro Matsumura
Graduate School of Economics
Osaka University
Toyonaka, Osaka, Japan

Katsutoshi Yada
Faculty of Business and Commerce
Kansai University
Osaka, Japan

Mitsunori Matsushita
Faculty of Informatics
Kansai University
Takatsuki, Osaka, Japan

Daisuke Katagami
Department of Applied Computer Science
Tokyo Polytechnic University
Atsugi, Kanagawa, Japan

Akinori Abe
Division of Behavioral Science
Chiba University
Chiba, Japan

Hisashi Kashima
Department of Intelligence Science
and Technology
Kyoto University
Kyoto, Japan

Toshihiro Hiraoka
Institute of Industrial Science
The University of Tokyo
Meguro-ku, Tokyo, Japan

Takahiro Uchiya
Department of Computer Science
and Engineering
Nagoya Institute of Technology
Nagoya, Japan

Rafal Rzepka
Graduate School of Information Science
and Technology Sapporo Hokkaido Japan
Hokkaido University
Sapporo, Hokkaido, Japan

ISSN 2194-5357 ISSN 2194-5365 (electronic)
Advances in Intelligent Systems and Computing
ISBN 978-3-030-96450-4 ISBN 978-3-030-96451-1 (eBook)
https://doi.org/10.1007/978-3-030-96451-1



This Springer imprint is published by the registered company Springer Nature Switzerland AG
The registered company address is: Gewerbestrasse 11, 6330 Cham, Switzerland

# Preface

This book contains extended versions of research papers presented at the international sessions at the 35th Annual Conference of the Japanese Society for Artificial Intelligence (JSAI2021), which was held online in June 2021. The JSAI annual conferences are considered key events for our organization, and the international sessions held at these conferences play a key role for the society in its efforts to share Japan's research on artificial intelligence with other countries. With the increased impact of AI on the world, the JSAI annual conferences have recently attracted many participants not only from academia but also from industries. JSAI2021 was held online, and various research ideas were shared through active discussion with many participants. We are extremely pleased to publish this collection of papers as the research results of our international sessions.

The topics of the international sessions cover five categories: knowledge engineering, machine learning, agents, robots and real worlds, and human interface and education aid, which have been important topics for the JSAI community. A total of 57 papers were selected for presentation at the international sessions, among which 35 papers submitted to a regular category were handled as the candidate papers for this post-proceedings. Each of the candidate papers was reviewed by at least two experts from the viewpoint of originality (the novelty of the paper), significance (impact on sciences and business), and quality of presentation. As a result of this review process, 25 papers were selected and invited to submit their extended version from papers presented at JSAI2021. After checking the quality of submitted extended versions of papers by JSAI2021 International Program Committee Members, 19 papers are included in this book. All of the papers are original and high quality, representing key contributions to AI research.

We would like to extend our deepest appreciation to President Itsuki Noda, Vice President Ryutaro Ichise (General Chair of JSAI2021), and Executive Committee Chair Takeo Hosomi, as well as to the JSAI administrative staff and Springer publishing staff for their tremendous assistance on this project. We would also like to thank the authors of the papers contained in this book and all international session contributors.

Yasufumi Takama
Chief Editor

Mitsunori Matsushita
Program Chair, JSAI 2021

Naohiro Matsumura
Vice Program Chair, JSAI 2021

Daisuke Katagami
Takahiro Uchiya
Akinori Abe
Katsutoshi Yada
Rafal Rzepka
Toshihiro Hiraoka
Hisashi Kashima
International Program Committee, JSAI 2021

# Organization

The 35th Annual Conference of the Japanese Society for Artificial Intelligence, 2021

## Conference Chair

| | |
|---|---|
| Ryutaro Ichise | National Institute of Informatics, Japan |

## Executive Committee Chairs

| | |
|---|---|
| Takeo Hosomi | NEC Corporation, Japan |
| Hiroyuki Toda (vice) | NTT Corporation, Japan |

## Program Committee Chairs

| | |
|---|---|
| Mitsunori Matsushita | Kansai University, Japan |
| Naohiro Matsumura (vice) | Osaka University, Japan |
| Yoko Nishihara (sub) | Ritsumeikan University, Japan |

## Program Committee

| | |
|---|---|
| Tomohiro Harada | Tokyo Metropolitan University, Japan |
| Satoshi Yoshida | NEC Corporation, Japan |
| Munehiko Sasajima | University of Hyogo, Japan |
| Eiichi Sunagawa | Toshiba Corporation, Japan |
| Koh Mitsuda | NTT Corporation, Japan |
| Daniel Andrade | NEC Corporation, Japan |
| Ryo Iwaki | IBM Research, Tokyo, Japan |
| Yuki Nakaguchi | NEC Corporation, Japan |
| Hidekazu Oiwa | Google, USA |
| Takashi Nishibayashi | Voyage Group, Japan |

# Contents

# Machine Reading Comprehension of News on Stock Price Changes

Seiji Suzuki, Gakuto Tsutsumi, and Takehito Utsuro(✉)

Graduate School of Science and Technology, University of Tsukuba, 1-1-1, Tennodai, Tsukuba, Ibaraki 305-8573, Japan
utsuro@iit.tsukuba.ac.jp

**Abstract.** One of the reasons why stock prices fluctuate greatly is because of IR announcements, which are information for investors, and news reports on events that are closely related to the companies. When a stock price change occurs, news sites for investors may report the stock price change and the reason for the change. However, such articles report only a certain portion of overall events that are closely related to reasons for stock price changes. Thus, in order to provide investors with information on those reasons for stock price changes, it is necessary to develop a system to collect information on events that could be closely related to the stock price changes of certain companies from the Internet. As the first step towards developing such a system, this paper takes an approach of employing a BERT-based machine reading comprehension model, which extracts causes of stock price changes from news reports on stock price changes. Those extracted reasons are intended to be further used to train a system to collect information on events that could be closely related to the stock price changes of certain companies from the Internet.

## 1 Introduction

One of the reasons why stock prices fluctuate greatly is because of IR announcements, which are information for investors, and news reports on events that are closely related to the companies. When such information is delivered, as shown in Fig. 1, the trading volume of the companies increase and the a stock price change occurs. When a stock price change occurs, news sites for investors may report the stock price change and the reason for the change as shown in Fig. 2. However, such articles report only a certain portion of overall events that are closely related to reasons for stock price changes. Thus, in order to provide investors with information on those reasons for stock price changes, it is necessary to develop a system to collect information on events that could be closely related to the stock price changes of certain companies from the Internet. As the first step towards developing such a system, this paper takes an approach of employing a BERT [5]-based machine reading comprehension model [14], which extracts reasons for stock price changes from news reports on stock price changes


Y. Takama et al. (Eds.): JSAI 2021, AISC 1423, pp. 1–12, 2022.
https://doi.org/10.1007/978-3-030-96451-1_1

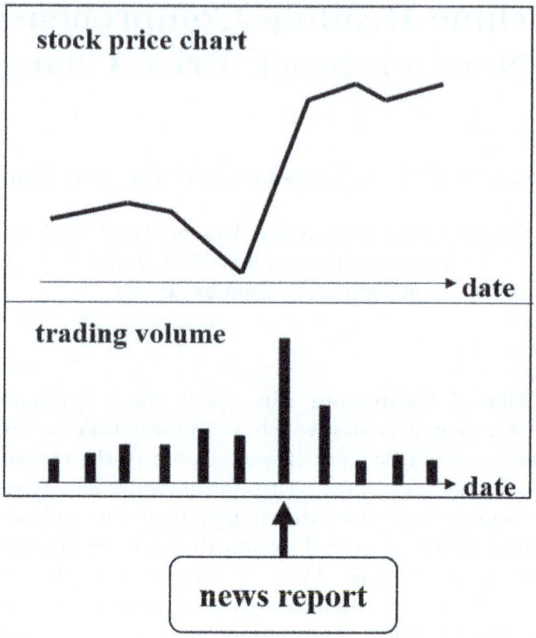


**Fig. 1.** Relation of stock price changes and trading volume per day and news report

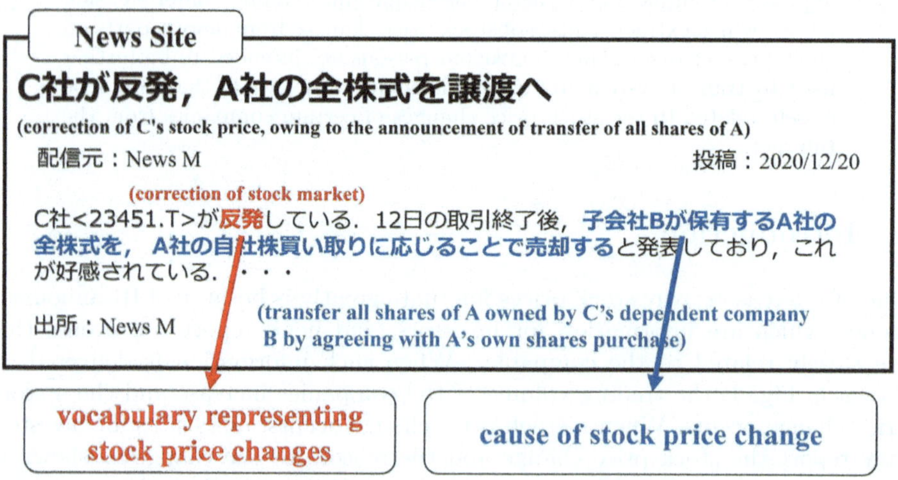


**Fig. 2.** The vocabulary representing stock price changes and a cause of stock price change: an example

(Fig. 3). Those extracted reasons are intended to be further used to train a system to collect information on events that could be closely related to the stock price changes of certain companies from the Internet.

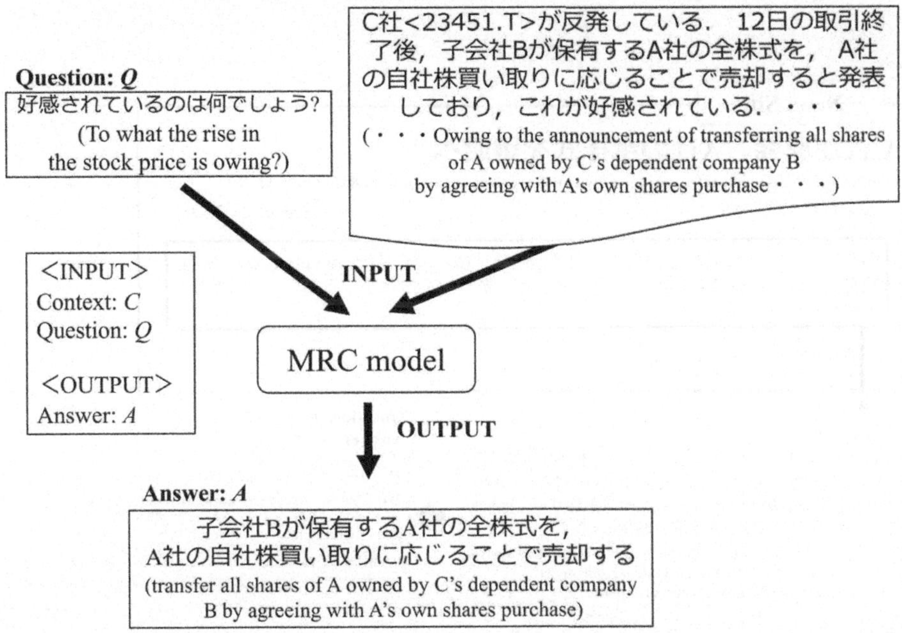


**Fig. 3.** The framework of machine reading comprehension of causes of stock price changes

The task of machine reading comprehension is formalized as below: given a question and a context text where the answer to the question is stated, the task is to extract the answer from the context text. The task of machine reading comprehension is divided into factoid and non-factoid machine reading comprehension sub-tasks. In factoid machine reading comprehension, it is known that the performance by a machine is over that by a human[1]. The task of machine reading comprehension of causes of stock price changes is regarded as a non-factoid machine reading comprehension sub-task. Out of such non-factoid machine reading comprehension sub-tasks, Chen et al. [4] studied the task of how-to tip machine reading comprehension. Chen et al. [4] explored a way to develop a dataset for training Japanese how-to tip QA models, and it applied BERT [5] to a Japanese how-to tip QA dataset. Evaluation results of Chen et al. [4] revealed that the how-to tip machine reading comprehension performance was almost comparative with that of the factoid machine reading comprehension even with the training data size reduced to around 4% of the factoid machine reading comprehension.

[1] When evaluated with the SQuAD [14] (https://rajpurkar.github.io/SQuAD-explorer/).

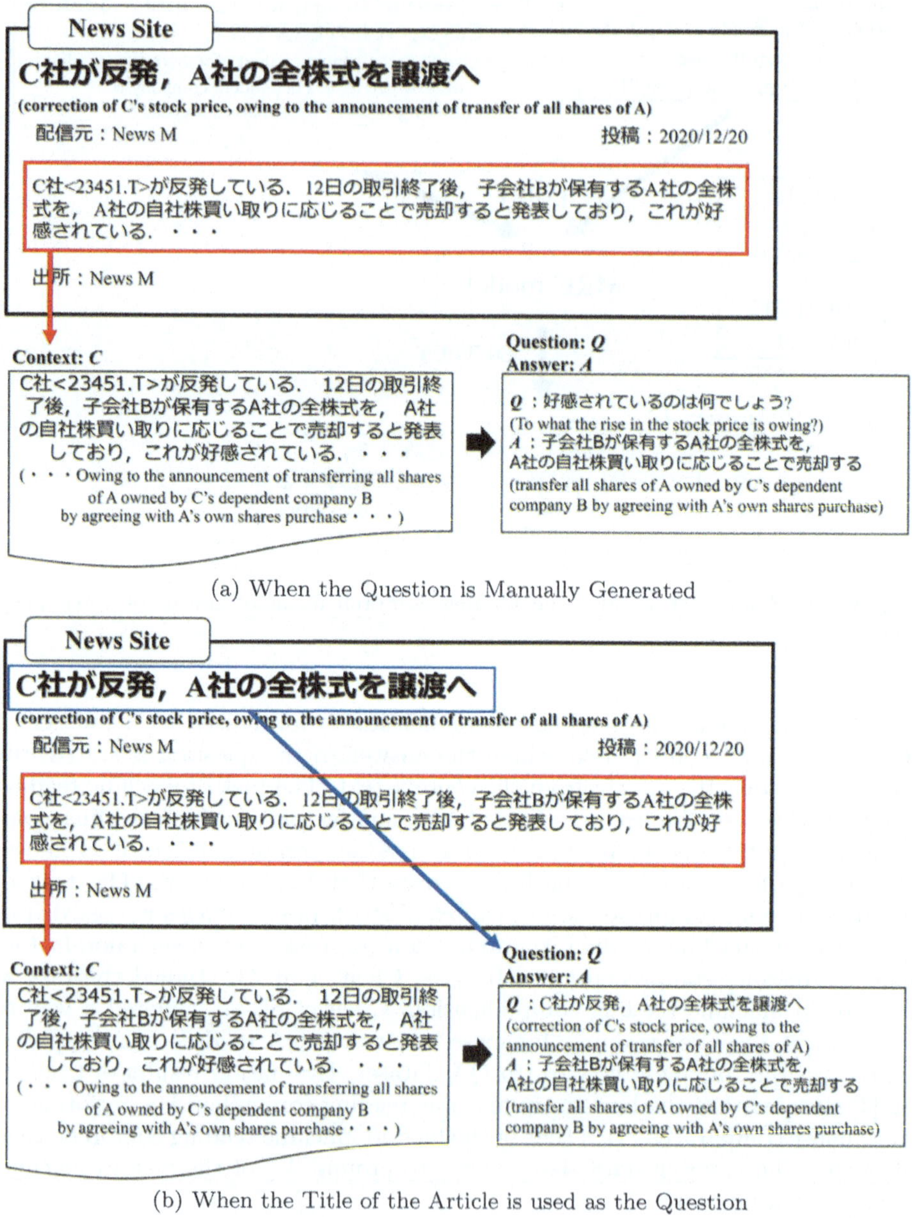


**Fig. 4.** The procedure of developing an example of machine reading comprehension of causes of stock price changes

**Table 1.** Statistics of the Categories of 100 Articles delivered from "みんなの株式 (minna-no kabu-shiki)"

| Category | # of articles |
|---|---|
| News on stock price changes and their causes | 25 |
| News on companies such as the announcements on New products | 7 |
| News on domestic equities | 24 |
| News on foreign equities | 9 |
| News on exchange market | 7 |
| News on bond market | 2 |
| News on financial market | 2 |
| News for individual investors | 21 |
| Economic indicator | 3 |
| Total | 100 |

## 2 Stock Price News of "みんなの株式 (minna-no kabu-shiki)"

In this paper, the news site from which we collect the stock price news is minkabu.jp[2], where we used 3,300 articles[3] delivered from "みんなの株式 (minna-no kabu-shiki)." minkabu.jp includes about 260,000 articles (as of January 2021) that are delivered from "みんなの株式 (minna-no kabu-shiki)". Table 1 shows the statistics of the categories of 100 articles randomly sampled from the collected 3,300 articles. Based on this statistics, out of the overall 260,000 articles delivered from "みんなの株式 (minna-no kabu-shiki)", it is estimated that the number of articles on "news on stock price changes and their causes" amount to 65,000 (25%). Thus, "みんなの株式 (minna-no kabu-shiki)" can be considered as a resource with a sufficient number of articles for developing a dataset for the examples of machine reading comprehension of causes of stock price changes.

Out of the "rise" and "decline" of stock prices, this paper focuses on "rise" of stock prices and develop a dataset of examples of machine reading comprehension of causes of stock price changes. First, we collect 11 (and several more) vocabularies listed in Table 2 representing "rise" in stock prices and randomly selected 627 articles that include one of those "rise" vocabularies, out of the overall 3,300. This is the result of discarding 14 articles that are not appropriate for developing examples of machine reading comprehension of causes of stock price changes. This is also the result of discarding 35 articles including another 13 vocabularies representing "decline" in stock prices. Those 13 vocabularies include

[2] https://minkabu.jp/.
[3] Delivered from October 30th to December 3rd, 2020.

"反落 (reactionary fall)", "下落 (decline)", "続落 (continued fall)", "急落 (fall rapidly)", "売りに押され (drop)", and "安値 (low price)".

From those 627 articles, we develop the dataset of examples of machine reading comprehension of causes of stock price changes.

**Table 2.** # of Occurrences and their Ratio (%) of Vocabularies representing Rise in Stock Prices among 627 examples of machine reading comprehension of causes of stock price changes

| Vocabulary | # | Ratio |
|---|---|---|
| 反発 (correction) | 176 | 19.7 |
| 続伸 (continued rise) | 172 | 19.3 |
| 高値 (high price) | 115 | 12.9 |
| カイ気配 (bid price) | 87 | 9.7 |
| 大幅高 (large rise) | 66 | 7.4 |
| 上昇 (rise) | 56 | 6.3 |
| ストップ高 (hit limit high) | 54 | 6.0 |
| 急伸 (rise rapidly) | 49 | 5.5 |
| 連騰 (winning streak) | 40 | 4.5 |
| 堅調 (increase steadily) | 38 | 4.3 |
| 急騰 (sharp rise) | 35 | 3.9 |
| other | 5 | 0.5 |
| total | 893 | 100 |

## 3 Generating Examples of Machine Reading Comprehension of Causes of Stock Price Changes from Stock Price News Articles

In the procedure of generating examples of machine reading comprehension of causes of stock price changes from stock price news articles, we take the following two approaches: (a) The question manually generated from the article is used as the question $Q$ of the machine reading comprehension. (b) The title of the article is used as the question $Q$ of the machine reading comprehension. In both cases, the whole article is used as the context $C$ of the machine reading comprehension.

(a) This case is described in Fig. 4(a). The question manually generated from the article is used as the question $Q$ of the machine reading comprehension.
(b) This case is described in Fig. 4(b). The title of the article is used as the question $Q$ of the machine reading comprehension.

**Table 3.** # and Ratio of Manually Generated Questions $Q$ among 627 Examples of Machine Reading Comprehension of Causes of Stock Price Changes

| Question $Q$ | # | ratio |
|---|---|---|
| 材料視されているのは何でしょう?<br>(What is the good offer in front of the market?) | 130 | 20.7 |
| 好感されているのは何でしょう?<br>(To what the rise in the stock price is owing?) | 122 | 19.5 |
| 寄与したのは何でしょう?<br>(What contributes to the rise in the stock price?) | 61 | 9.7 |
| 要因は何でしょう?<br>(What causes the rise in stock prices?) | 48 | 7.7 |
| 発表したのは何でしょう?<br>(What is announced?) | 38 | 6.1 |
| 好調なのは何でしょう?<br>(What is fare well?) | 29 | 4.6 |
| 利益を押し上げたのは何でしょう?<br>(What boosted the profits?) | 25 | 4.0 |
| 期待されているのは何でしょう?<br>(What is expected regarding the rise in the stock price?) | 14 | 2.2 |
| 背景は何でしょう?<br>(What is the background of the rise in the stock price?) | 12 | 1.9 |
| other (each # $\leq$ 10) | 148 | 23.6 |
| 合計 | 627 | 100 |

Within sentences including the answer span, a vocabulary indicating the "cause" of the rise of the stock price (henceforth referred to as the "cause" vocabulary) usually appears. The overall number of those "cause" vocabularies is 62, which include the followings:

"材料視" (good offer in front of the market), "好感" (owing to), "期待" (expect), "発表" (announce), "株価を刺激" (motivates the rise in stock prices), "要因" (cause), "引き上げ" (raised), and "追い風" (positive).

The case (a) is quite straightforward in that we manually generate question sentences from those "cause" vocabularies as shown in the statistics of Table 3[4]. When beyond the machine reading comprehension setting, where only the question $Q$ is available and the candidates of context $C$ have to be automatically collected from a large pool of documents [3,9], we first identify those "cause"

[4] As an alternative, we also examine the approach of using a single general question sentence such as below:「A 社の株価が変動した理由はなんでしょう？」However, this approach does not perform well in the evaluation compared with the case (a) in this section.

vocabularies within the collected candidates of contexts, then the question $Q$ is automatically switched according to the identified "cause" vocabulary and then the machine reading comprehension model is applied to the tuple of the question $Q$ and the context $C$. The case (b), on the other hand, has an advantage in that we can omit the procedure of manually generating the question from the article.

# 4 Evaluation

## 4.1 Evaluation Procedure

As the version of BERT [5] implementation which can handle a text in Japanese, the TensorFlow version[5] and the Multilingual Cased model[6] were used as the pre-trained model. Before applying BERT modules, MeCab[7] was applied with mecab-ipadic-NEologd dictionary[8] and the Japanese text was segmented into a morpheme sequence. Then, within the BERT fine-tuning module, the WordPiece module with 110k shared WordPiece vocabulary was applied, and the Japanese text was further segmented into a subword unit sequence. Finally, the BERT fine-tuning module for machine reading comprehension[9] was applied as well as the fine-tuned model. In the fine-tuning procedure, the BERT pre-trained model was first fine-tuned with 12,000 factoid machine reading comprehension training examples[10], and then, the fine-tuned model was further fine-tuned with the training examples of machine reading comprehension of causes of stock price changes developed in the previous section.

## 4.2 Evaluation Results

In the evaluation, we examined the following three cases:

(a) This case is described in Fig. 4(a) and as the case (a) in Sect. 3.
(b) This case is described in Fig. 4(b) and as the case (b) in Sect. 3.
(c) The question $Q$ is the same as the case (a). The title of the article is concatenated to the context of the cases (a) and (b) and then used as the context $C$.

Figure 5 shows the evaluation results of 5-fold cross validation. Although the case (b) is beneficial in that it can avoid the manual procedure of generating a question from the article, the best performing ones are the case (a) or the case (c), where the performance of the case (b) is about 76% for the exact match,

[5] https://github.com/google-research/bert.
[6] Trained with 104 languages, available from https://github.com/google-research/bert/blob/master/multilingual.md.
[7] http://taku910.github.io/mecab/ (in Japanese).
[8] https://github.com/neologd/mecab-ipadic-neologd.
[9] run_squad.py, with the number of epochs as 2, batch size as 8, and learning rate as 0.00003.
[10] http://www.cl.ecei.tohoku.ac.jp/rcqa/ (in Japanese).

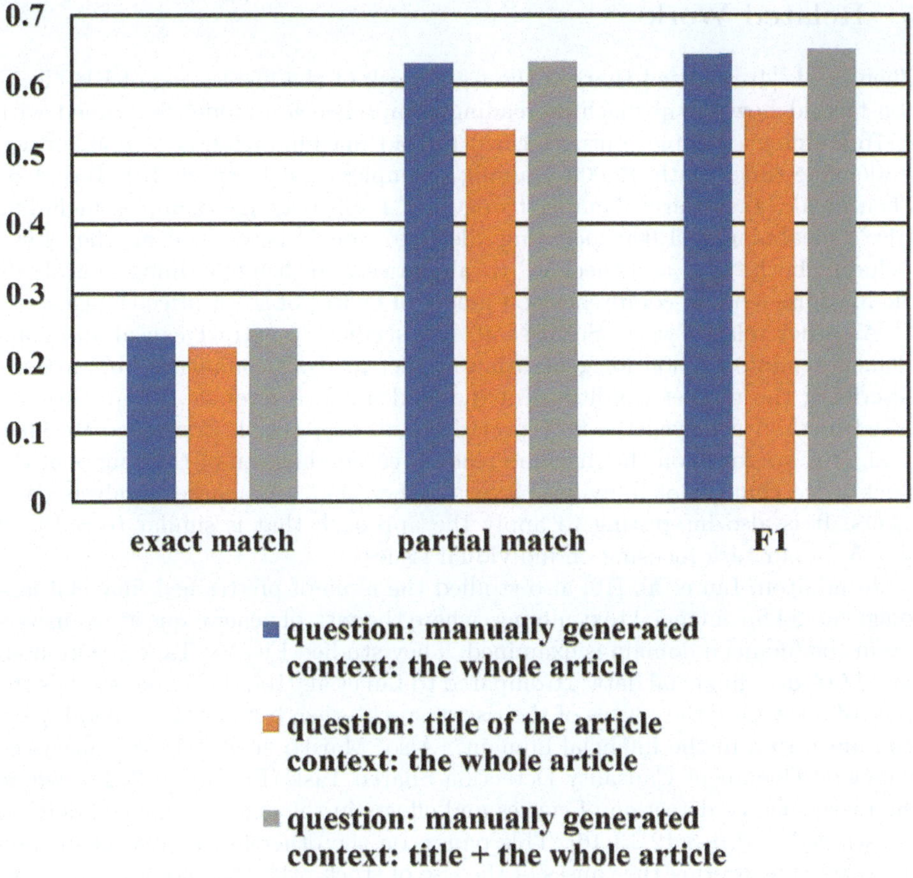


**Fig. 5.** Evaluation results

53% for the partial match, and 78% for the F1 score. For the case (b), we also examine the difference between the title of the article and the answer span. Then, out of the overall 627 examples for evaluation, for 496 examples (about 80 %), answer span does include additional information that are not included in the title. So, this means that for some of those 80% cases, hopefully the model of the case (b) successfully identifies some of those additional answer spans that are not included in the title of the article without manually creating the questions. Actually, as the result of manually examining the exact and partial matches of the case (b), for about 70–90% of the exact and partial matches, we observe that the model successfully identify those additional answer spans that are not included in the question.

## 5 Related Work

Chen et al. [4] reported that, in the framework of the fine-tuning of BERT [5], non-factoid how-to tip machine reading comprehension model fine-tuned with 2,576 training examples outperformed factoid machine reading comprehension model fine-tuned with 12,000 training examples (see footnote 10). Based on Chen et al. [4], we are planning to develop 1,000 training examples including "rise" vocabularies, 1,000 including "decline" vocabularies, and another 1,000 including both "rise" and "decline" vocabularies, and then fine-tune and evaluate the machine reading comprehension model of causes of stock price changes.

As other related work, Sakai et al. [15] studied to extract causal and complementary information for generating market analysis comments from articles describing the market conditions of the Nikkei Stock Average, where the proposed method achieved the 90% precision and 61% recall. Compared to Sakai et al. [15], we focus on the machine reading comprehension of the causes of the stock price changes of individual issues rather than the market analysis comments. It is also interesting to apply the approach that is similar to Sakai et al. [15] in our task focusing on individual issues.

In additon, Liu et al. [10] also studied the issue of pre-trained financial language model for financial text mining, where the task of general question answering in the financial domain is examined. They studied FiQA[11] Task 2 "Opinion-based QA over financial data". Compared to Liu et al. [10], this paper studies the issue of answering the causes of the rise of stock price, but not the general question answering in the financial domain. Also, Mariko et al. [11] organized the Financial Document Causality Detection Shared Task (FinCausal 2020), where the tasks such as detection of causes and effects in the general financial domain are studied [1,2,6–8,12,13,16]. This paper, on the other hand, concentrates on the issue of answering the causes of the rise of stock price, but not on the causes and effects in the general financial domain.

## 6 Conclusion

This paper took an approach of employing a BERT [5]-based machine reading comprehension model [14], which extracts reasons for stock price changes from news reports on stock price changes. Overall, the approach of using the title of the article as the question $Q$ of the machine reading comprehension is recommended in that the procedure of manually generating the question from the article can be omitted, although its performance is slightly lower than the approach of manually creating questions. Future work includes incorporating not only the "rise" of stock prices but also the "decline" of them. Another future work is to increasing the number of training examples and to examine whether those increased training examples contribute to improving the machine reading comprehension performance. It also includes scaling up into beyond the machine reading comprehension setting where only the question $Q$ is available and the

[11] https://sites.google.com/view/fiqa/home.

candidates of context $C$ have to be automatically collected from a large pool of documents [3,9].

## References


1. Becquin, G.: GBe at FinCausal 2020, task 2: span-based causality extraction for financial documents. In: Proceedings of 1st Joint Workshop on Financial Narrative Processing and MultiLing Financial Summarisation, pp. 40–44 (2020)
2. Chakravarthy, S., Kanakagiri, T., Radhakrishnan, K., Umapathy, A.: Domino at FinCausal 2020, task 1 and 2: causal extraction system. In: Proceedings of 1st Joint Workshop on Financial Narrative Processing and MultiLing Financial Summarisation, pp. 90–94 (2020)
3. Chen, D., Fisch, A., Weston, J., Bordes, A.: Reading Wikipedia to answer open-domain questions. In: Proceedings of 55th ACL, pp. 1870–1879 (2017)
4. Chen, T., Li, H., Kasamatsu, M., Utsuro, T., Kawada, Y.: Developing a how-to tip machine comprehension dataset and its evaluation in machine comprehension by BERT. In: Proceedings of 3rd FEVER, pp. 26–35 (2020)
5. Devlin, J., Chang, M.W., Lee, K., Toutanova, K.: BERT: pre-training of deep bidirectional Transformers for language understanding. In: Proceedings of NAACL-HLT, pp. 4171–4186 (2019)
6. Imoto, T., Ito, T.: JDD @ FinCausal 2020, task 2: financial document causality detection. In: Proceedings of 1st Joint Workshop on Financial Narrative Processing and MultiLing Financial Summarisation, pp. 50–54 (2020)
7. Ionescu, M., Avram, A.M., Dima, G.A., Cercel, D.C., Dascalu, M.: UPB at FinCausal-2020, tasks 1 & 2: causality analysis in financial documents using pre-trained language models. In: Proceedings of 1st Joint Workshop on Financial Narrative Processing and MultiLing Financial Summarisation, pp. 55–59 (2020)
8. Kao, P.W., Chen, C.C., Huang, H.H., Chen, H.H.: NTUNLPL at FinCausal 2020, task 2: improving causality detection using Viterbi decoder. In: Proceedings of 1st Joint Workshop on Financial Narrative Processing and MultiLing Financial Summarisation, pp. 69–73 (2020)
9. Lee, K., Chang, M.W., Toutanova, K.: Latent retrieval for weakly supervised open domain question answering. In: Proceedings of 57th ACL, pp. 6086–6096 (2019)
10. Liu, Z., Huang, D., Huang, K., Li, Z., Zhao, J.: FinBERT: a pre-trained financial language representation model for financial text mining. In: Proceedings of 29th IJCAI, pp. 4513–4519 (2020)
11. Mariko, D., Abi-Akl, H., Labidurie, E., Durfort, S., De Mazancourt, H., El-Haj, M.: The financial document causality detection shared task (FinCausal 2020). In: Proceedings of 1st Joint Workshop on Financial Narrative Processing and MultiLing Financial Summarisation, pp. 23–32 (2020)
12. Özenir, G., Karadeniz İ.: ISIKUN at the FinCausal 2020: linguistically informed machine-learning approach for causality identification in financial documents. In: Proceedings of the 1st Joint Workshop on Financial Narrative Processing and MultiLing Financial Summarisation, pp. 85-89 (2020)
13. Pielka, M., et al.: Fraunhofer IAIS at FinCausal 2020, tasks 1 & 2: using ensemble methods and sequence tagging to detect causality in financial documents. In: Proceedings of 1st Joint Workshop on Financial Narrative Processing and MultiLing Financial Summarisation, pp. 64–68 (2020)

14. Pranav, R., Jian, Z., Konstantin, L., Percy, L.: SQuAD: 100,000+ questions for machine comprehension of text. In: Proceedings of EMNLP, pp. 2383–2392 (2016)
15. Sakai, H., Sakaji, H., Izumi, K., Matsui, H., Irie, K.: Extraction of causal and complementary information for generating market analysis comments by automatic generation of training data. In: Proceedings of 34th Annual Conference JSAI (2020). (in Japanese)
16. Szántó, Z., Berend, G.: ProsperAMnet at FinCausal 2020, task 1 & 2: modeling causality in financial texts using multi-headed transformers. In: Proceedings of 1st Joint Workshop on Financial Narrative Processing and MultiLing Financial Summarisation, pp. 80–84 (2020)

# A Deep Fusion Model Combining News Content and Historical Prices for Stock Trend Prediction

Jeng-Haur Wang(✉) and Hsin-Wen Liu

National Taipei University of Technology, Taipei, Taiwan
jhwang@ntut.edu.tw

**Abstract.** Stock markets are often influenced by various factors which makes it very challenging to predict. Machine learning and deep learning models are often used to predict stock trends from its historical prices. Since there's a lot of online information available in addition to stock prices, including technical indicators, news reports, and social information, we intend to combine news content for improving the performance. In this paper, we propose a deep fusion model for stock trend prediction combining news content with historical stock prices. Firstly, we utilize multi-layered Long Short-Term Memory (LSTM) to learn sequential information from stock prices. Then, we adopt Hybrid Attention Networks (HAN) which include both sentence-level and temporal attention to discover the relative importance of words from news reports. Finally, we compare early and late fusion models to improve stock trend prediction. The experimental results show that the best macro-F1 score of 79.0% can be achieved when we use late fusion to aggregate the prediction results of news content using 2-layer LSTM and that of historical prices in a 5-day window using HAN. As compared to individual models, the performance improvement of up to 40% can be obtained. This shows the potential of our proposed approach. Further investigation is needed for stock trend prediction in different markets.

**Keywords:** Stock trend prediction · Deep learning · Semantic analysis

## 1 Introduction

Stock markets are usually influenced by many different factors, including company revenues, industry growth, global economics, and political situations, to name a few. This makes it very challenging to predict the trend. Conventionally, it's common to conduct technical analysis based on stock prices and basic analysis based on company revenues for stock trend prediction in the industry. Machine learning and deep learning models are often used to predict stock trends given huge amount of data. Nowadays, there's a lot more online information available in addition to stock prices and technical indicators, including company revenues, news reports, and social information. Among these types of information, we intend to explore the effects of news content on stock trend prediction, since news reports might contain useful insights into the developments and perspectives of the corresponding company or industry. Given the rapid changes


Y. Takama et al. (Eds.): JSAI 2021, AISC 1423, pp. 13–26, 2022.
https://doi.org/10.1007/978-3-030-96451-1_2

and global impact from different markets and industries, it would be useful to extract important semantic features for more accurate prediction.

In this paper, we propose a deep fusion model for stock trend prediction combining news content with historical stock prices. Firstly, we utilize multi-layer Long Short-Term Memory (LSTM) to learn sequential information from stock prices and trading volumes. Then, we adopt Hybrid Attention Networks (HAN) [10] which include both sentence-level and temporal attention to discover the relative importance of words from news reports. Finally, early and late fusion models are compared to explore the best way to aggregate features for improving prediction performance. From our experimental results on S&P500, the best macro-F1 score of 79.0% can be achieved when we use late fusion to combine prediction results of news content using 2-layer LSTM and that of stock prices in a 5-day window using HAN. As compared to individual models, the performance improvement of up to 40% can be obtained. This shows the potential of our proposed approach. Further investigation is needed to verify the performance for stock trend prediction in different markets.

## 2 Related Work

Early studies on stock trend prediction utilize regression or time-series models, such as ARIMA. Due to the availability of more and more online information in addition to historical stock prices, such as and news articles, and social information, recent research on stock trend prediction focuses on machine learning and deep learning methods based on individual information. For methods based on historical stock prices, Meesad et al. [1] proposed to predict stock prices using support vector regression (SVR). Kumar et al. [2] compared five different machine learning methods for stock prediction, and found Naïve Bayes classifiers outperform others types. Hiransha et al. [3] used deep learning models including Multi-Layer Perceptrons (MLPs), CNNs, RNNs, and LSTMs, and found them superior to time series linear models. Chen et al. [4] divided the ranges of stock returns into 7 categories, and predicted using LSTM with the accuracy of 27.2%. Adebiyi et al. [5] found ANN models to perform better than ARIMA. Gao et al. [6] used the opening, closing, and volume to predict the closing price next day, in which LSTM performs better. Nivetha et al. [7] also confirmed the better performance of deep learning models than conventional MLP and SVM.

On the other hand, Lee et al. [8] found text analysis to be an important factor for stock prediction. Thus, there are also many methods based on news articles. For example, Ding et al. [9] proposed to train event embedding using news titles and predict the volatility using CNNs. Hu et al. [10] predicted the volatility of stock prices using bidirectional GRU models with the accuracy of 47%. Akita et al. [11] trained a deep learning model for stock price prediction. Shi et al. [12] predicted stock prices using texts. Pagolu et al. [14] found strong correlation between the public sentiments and stock prices. Sehgal et al. [15] confirmed the high correlation between stock prices and the sentiments. In addition, Li et al. [13] found the combination of stock prices and news can improve the accuracy of prediction. Thus, there are some methods based on both stock prices and news articles. For example, Zhai et al. [16] combined news and technical indicators to improve the accuracy of prediction. Li et al. [17] combined news sentiments, technical indicators, and stock prices using LSTM to obtain better performance. Deng et al. [18] found the combination of time series data with social information to outperform other methods when predicting stock prices. Xu et al. [19] proposed a deep generative model to predict stock prices. In this paper, we propose a deep fusion model for stock trend prediction by combining features learned from news content and historical stock prices by two different deep learning models.

## 3 The Proposed Method

In this paper, we propose a deep fusion model for stock prices prediction using two types of data: historical prices, and news articles. There are three modules in the system architecture: preprocessing, feature extraction and fusion, and stock trend prediction, as shown in Fig. 1.

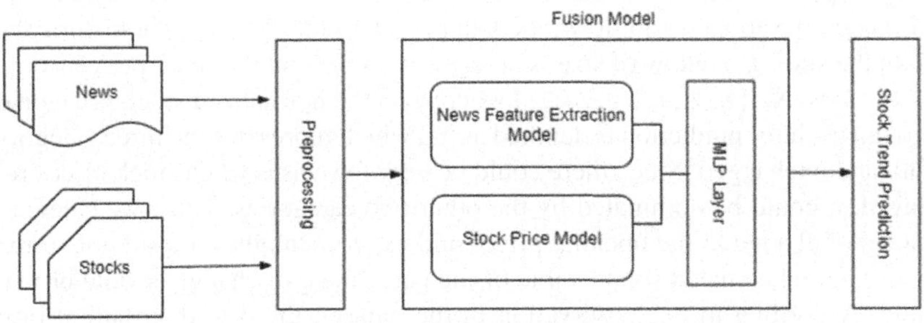


**Fig. 1.** The system architecture of the proposed approach

As shown in Fig. 1, the historical stock prices are collected as well as the news articles during the same period of time, and aligned in the preprocessing step. In order to learn the characteristics of time series data of stock prices, we adopt a multi-layer LSTM as the major model which can deal with the long-term dependencies with the corresponding memory cells. To avoid over-fitting, we include an extra dropout layer after each LSTM layer. Then, to extract useful semantic features from news articles for stock price prediction, we adopt a hybrid attention network model [10] which combines news-level and temporal attention to address the sequential context information among news. Finally, we further aggregate the results of these two models with early and late fusion models to predict the trend of stock prices. The details of each module are explained in the following subsections.

### 3.1 Preprocessing

To collect historical stock prices and the corresponding news articles, we choose our target as the 500 companies in S&P 500 index since they are the most representative and traded in the market. Firstly, we utilize Financial Data APIs from Quandl to obtain the opening, closing, high, and low prices, as well as the trading volume for each of the 500 companies each day. Then, the corresponding news articles in the same days are collected from Reuters. For the preprocessing of historical stock prices, we first apply z-score normalization to lower the effect of outliers. This can improve the convergence and accuracy of the model.

For the preprocessing of news articles, we first filter out news without mentioning any of the 500 companies, and ordered by date. Then, we reassign news articles on Saturdays and Sundays to Friday, since the markets are closed on weekends. This aligns all news articles to the corresponding stock trading days so that we can better investigate their relations in a meaningful way.

## 3.2 Feature Extraction

After aligning the news articles to the corresponding stock by date, we need to extract the major features as input to the deep learning models. Since the factors affecting stock prices might be very diverse, in this paper, we assume that the stock price is influenced by a certain period of time ahead, for example, five or ten days. Thus, as the input of LSTM model, we adopt a sliding window approach to representing the historical stock prices. In the sliding window of size N at day i, we represent the stock prices from day i to $i + N - 1$ as: $X_i$: $[z_i, z_{i+1}, \ldots z_{i+N-1}]$ where $z_i$ is the normalized stock price on day i. The ground truth for prediction is denoted as $Y_i$, which represents the three categories of rise, fall, and unchanged. Since there could be very few cases when stock prices remain unchanged, it could be dominated by the other two categories. Thus, we calculate the percentage of change in the opening prices and experimentally set a suitable threshold for unchanged when the absolute value of the percentage of change is smaller than the threshold. According to our observation in the dataset, the data distribution in terms of the ratios of rise, fall, and unchanged using different thresholds are shown in the following table. Considering a more even data distribution among three categories, the threshold is set to 0.45% (Table 1).

**Table 1.** Effects of different thresholds on the ratios of rise, fall, and unchanged

| Threshold (%) | Rise | Fall | Unchanged |
|---|---|---|---|
| 0.35 | 0.22 | 0.37 | 0.41 |
| 0.4 | 0.28 | 0.34 | 0.38 |
| 0.45 | **0.35** | **0.31** | **0.34** |
| 0.5 | 0.39 | 0.22 | 0.39 |

In the representation of news texts, we adopt the PV-DM model from the document embedding Doc2Vec [21] instead of word embedding Word2Vec [22]. Doc2Vec was shown by Lau et al. [20] to have better performance than Word2Vec [22]. There are two types of paragraph vectors: PV-DM (Distributed Memory Model of Paragraph Vectors), and PV-DBOW (Distributed Bag-of-Words version of Paragraph Vectors). Since PV-DBOW is based on bag-of-words, word order was not preserved. Thus, we choose the PV-DM model. Each document is represented as a 200-dimension vector. In accordance with the sliding window approach for stock price representation, we use the same window size as the number of trading days. To address the semantic meaning from daily news in the Hybrid Attention Network model, we include a limit on the maximum number of news per day for a more balanced representation each day.

## 3.3 Deep Learning Models

Stock prices of a given company is a kind of time series data. The stock price on a particular day "continues" from the previous trading day in the sense that the opening price is generally determined from the market based on the closing price on the previous trading day. Although there are many factors that could affect the trading behavior on the new day, thus the opening price, there's still a limit to the possible range for the high and low. To learn the variation in such a sequence, we utilize a multi-layer LSTM as our model.

**Stock Trend Learning Using Multi-layer LSTM.** Recurrent Neural Networks (RNNs) are designed to learn characteristics from time series data. Due to the potential vanishing gradients problem in a long sequence, Long Short-Term Memory (LSTM) models are designed to learn long-term dependency in the sequence.

In this paper, the inputs to the Multi-layer LSTM include the opening, closing, high, and low prices, as well as the trading volume of the stock for each company. The system architecture of a Multi-layer LSTM is shown in Fig. 2.

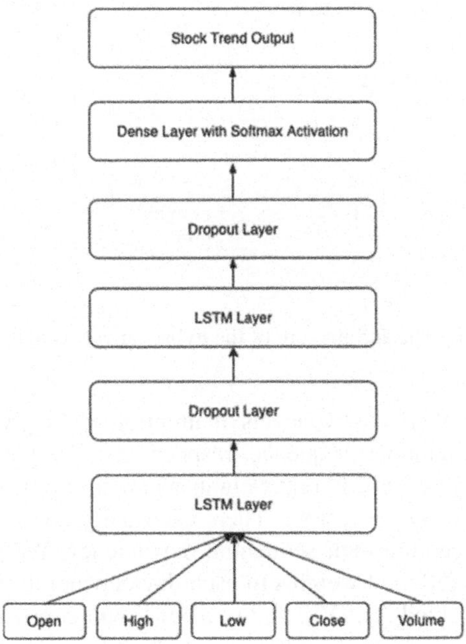


**Fig. 2.** The architecture of a multi-layer LSTM for stock trend prediction

As shown in Fig. 2, we first stack two LSTM layers to learn the sequential dependency from the inputs. Then, to avoid overfitting, each LSTM layer is followed by a dropout layer which randomly reset the weights of some edges to zero. Finally, a dense layer with three neurons is included which maps to the three output categories. The activation function of SoftMax is used to determine the corresponding probability for each category.

**News Semantic Learning Using Hybrid Attention Networks.** In text classification, Yang et al. [23] proposed a hierarchical attention network for extracting the word units in different levels, and assigning different importance weights. Financial news usually describe the current situations of the market, and also show the future perspectives of the companies and global economy. There could be different terms describing the various market phenomenon. For example, bull and bear markets are simple summary of possible rises and falls, respectively. In order to extract useful semantic features that reflect the economic situations, we utilize the hybrid attention networks [10] as a useful representation. The architecture is shown in Fig. 3.

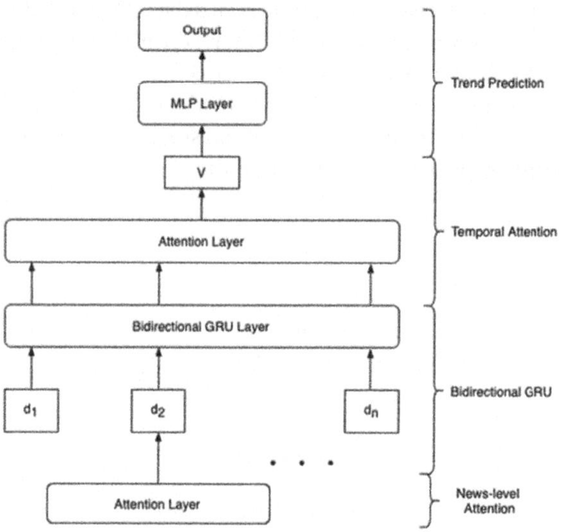


**Fig. 3.** The architecture of the hybrid attention networks.

As shown in Fig. 3, there are two levels of attention which give different importance weights for news and temporal sequence, respectively. Firstly, for each news article, news-level attention calculates different weights where the daily news vector is obtained by the weighted sum of news vectors. Then, each daily news vector di is input to a bidirectional GRU to capture their sequential dependency. Temporal attention is then applied which assigns different weights to each day depending on their relative importance. Finally, the weighted sum V is fed to a MLP which predicts the output as rise, fall, or unchanged.

### 3.4 Feature Fusion and Stock Trend Prediction

To combine features learned from news content and stock prices, we incorporate two fusion models: early fusion, and late fusion.

**Early Fusion.** In early fusion, we first concatenate each hidden state of the last LSTM layer in the multi-layer LSTM, and the output from bidirectional GRU in hybrid attention networks as an aggregated feature from historical stock prices and news semantics. Then, temporal attention is applied to calculate the attention weights by considering stock prices and news simultaneously. Finally, a MLP layer is used to predict the output. The idea of early fusion is shown in Fig. 4.

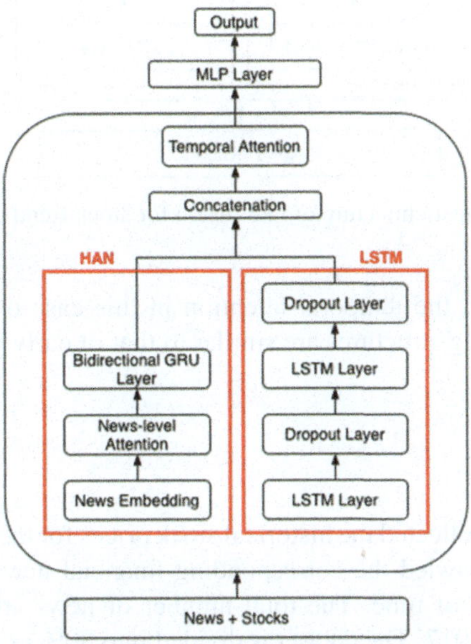


**Fig. 4.** The architecture of early fusion for stock trend prediction

As shown in Fig. 4, we can see the two models (LSTM and HAN) can run in parallel. In this case, the temporal attention from the original HAN is delayed until both models generate outputs, which are then concatenated as input to temporal attention.

**Late Fusion.** In late fusion, we wait for the complete models to be learned from both hybrid attention networks and multi-layer LSTM. Then, we concatenate the last hidden state from the last layer in multi-layer LSTM, and the output of temporal attention in HAN. Finally, a MLP layer is used to predict the output. The idea of late fusion is shown in Fig. 5.

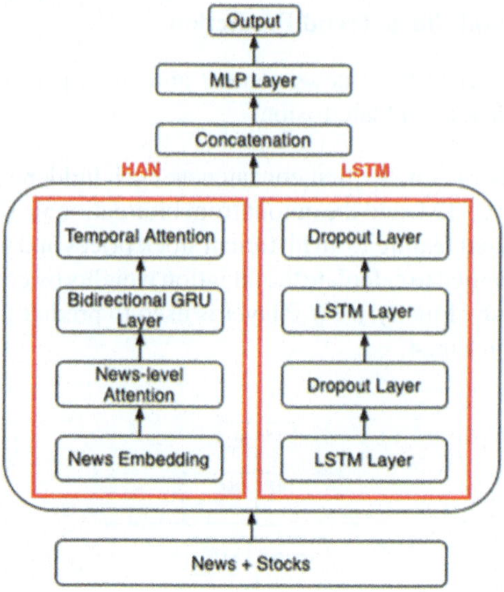


**Fig. 5.** The architecture of late fusion for stock trend prediction

As shown in Fig. 5, the temporal attention in this case only considers the news semantics. The remaining structures are similar to that of early fusion.

## 4 Experiments

In the experiments, we collected the historical stock prices for the 500 companies in S&P 500 index. Then, we crawled the corresponding financial news articles from Reuters during the same period of time. The total number of news articles is 10,686 during Jan.1, 2016 and Jul. 5, 2017. The dataset is divided into 80% as training set and 20% as test set. For the parameters of the deep learning models, we use Adam optimizer, and a learning rate of $5 * 10^{-5}$ with early stopping. Then, five-fold cross-validation is applied. The performance is evaluated using precision, recall, and F1-score.

Firstly, we checked the performance of individual models of stock trend prediction using historical prices, and using news articles, respectively. To verify the performance of the multi-layer LSTM on stock trend prediction, we first evaluated the effects of the number of layers on prediction performance. To avoid overfitting, we include a dropout layer after each LSTM layer. The results are shown as follows.

As shown in Table 2, we can observed the best performance when we used two layers in multi-layer LSTM. The best accuracy of 0.72 and F1 score of 0.7 can be obtained. In the remaining experiments, we set the number of layers as two.

Next, we evaluate the effects of the window size on the prediction performance, by setting the window size as 5 and 10. We also compared the different combinations of opening, closing, high, and low prices, as well as the trading volume as the features. The experimental result is shown in Table 3.

**Table 2.** Effects of the number of layers in Multi-layer LSTM for stock trend prediction

| # of layers | Accuracy | F1 Score |
|---|---|---|
| 1 | 0.64 | 0.58 |
| 2 | **0.72** | **0.7** |
| 3 | 0.69 | 0.65 |
| 4 | 0.65 | 0.59 |

**Table 3.** Effects of Multi-layer LSTM for stock trend prediction

| Window size = 5 | Close | Close/Volume | Close/Open/High/Low/Volume |
|---|---|---|---|
| Average accuracy | **0.72** | 0.46 | 0.65 |
| Average F1 score | **0.70** | 0.41 | 0.63 |
| Window size = 10 | Close | Close/Volume | Close/Open/High/Low/Volume |
| Average accuracy | 0.68 | 0.42 | 0.55 |
| Average F1 score | 0.56 | 0.39 | 0.51 |

As shown in Table 3, we can see the better performance when window size is equal to 5. The best performance can be obtained when we only used the historical closing prices in training, with an average accuracy of 0.72, and an average F1 score of 0.70. This shows stock prices might have higher variations in a longer period of time, which lowers the prediction accuracy.

Next, to evaluate the performance of hybrid attention networks for news semantics, we set the same window sizes of 5 and 10. The results are shown in Table 4.

**Table 4.** Effects of hybrid attention networks for stock trend prediction

| | Window size = 5 | Window size = 10 |
|---|---|---|
| Average accuracy | 0.4 | 0.4 |
| Average F1 score | 0.33 | 0.31 |

As shown in Table 4, we can see a similar performance for different window sizes. When window size is equal to 5, we can obtain a slightly better F1 score of 0.33. But, the general performance is not good when we only used information from news articles for prediction. This shows that news articles alone could not convey enough information to predict stock prices. Thus, in the following experiments, we further combined the features learned from news content and historical stock prices.

To observe the effects of combining news content with historical stock prices, we first evaluated the performance of early fusion as shown in Table 5.

**Table 5.** Effects of Early fusion

| Window size = 5 | Close | Close/Volume | Close/Open/High/Low/Volume |
|---|---|---|---|
| Average accuracy | **0.57** | 0.4 | 0.5 |
| Average F1 score | **0.51** | 0.37 | 0.46 |
| Window size = 10 | Close | Close/Volume | Close/Open/High/Low/Volume |
| Average accuracy | 0.44 | 0.38 | 0.39 |
| Average F1 score | 0.34 | 0.35 | 0.36 |

As shown in Table 5, we can observe the best performance when window size is equal to 5, using historical closing prices only. The best average accuracy is 0.57, and the average F1 score is 0.51. However, this is inferior to the original performance when using historical prices only as in Table 2. This might be due to the partial execution of HAN without temporal attention. When this feature was concatenated with the hidden state of multi-layer LSTM, the vector gets longer with the concatenation of LSTM outputs. But the news content are not successfully weighted by temporal attention.

Next, we further tested the performance of late fusion as in Table 6.

**Table 6.** Effects of late fusion

| Window size = 5 | Close | Close/Volume | Close/Open/High/Low/Volume |
|---|---|---|---|
| Average accuracy | **0.80** | 0.61 | 0.71 |
| Average F1 score | **0.79** | 0.57 | 0.69 |
| Window size = 10 | Close | Close/Volume | Close/Open/High/Low/Volume |
| Average accuracy | 0.77 | 0.47 | 0.62 |
| Average F1 score | 0.71 | 0.45 | 0.57 |

As shown in Table 6, we can observe the best performance when window size is equal to 5, using historical closing prices only. The best average accuracy is 0.80, and the average F1 score is 0.79. This is improved over the performance of using multi-layer LSTM alone. Since features from news content are fully learned using HAN with both sentence-level and temporal attention and concatenated with the LSTM outputs, the prediction performance can be improved by considering both features. This shows the benefits of aggregating news content with stock prices using late fusion.

To further justify the performance of our proposed approach, we compared with the method proposed by Li et al. [17]. In their method, they first concatenated historical stock prices and news content. Then, LSTM is used to predict stock prices. The average accuracy of 0.73 and an average F1 score of 0.69 can be obtained. This is inferior to our proposed approach. It shows the advantage of our proposed approach, especially for the multi-layer design for LSTM and late fusion. It also verifies the point that different types of data need different deep learning methods to extract the most important features for prediction.

Finally, we further explore the impact of news content and historical stock prices on stock price prediction in individual companies. In this paper, we selected three largest companies from four major sectors in the industry indexed by S&P 500 (Table 7).

**Table 7.** Data about the 12 representative companies

| Company name | Sector | Mean | Stddev | Max |
|---|---|---|---|---|
| Boeing | Industrials | 148.35 | 23.93 | 202 |
| American Airlines | Industrials | 40.81 | 5.74 | 51 |
| 3M | Industrials | 176.3 | 15.53 | 213 |
| Apple | Information technology | 116.34 | 18.81 | 156 |
| IBM | Information technology | 155.73 | 13.69 | 181 |
| Microsoft | Information technology | 58.98 | 6.32 | 72 |
| JPMorgan Chase | Financials | 72.94 | 12.13 | 93 |
| Bank of America | Financials | 18.16 | 4.45 | 25 |
| Prudential Financial | Financials | 88.6 | 15.94 | 113 |
| Amazon | Consumer discretionary | 763.9 | 125.82 | 1011 |
| General Motors | Consumer discretionary | 32.64 | 2.61 | 38 |
| Nike | Consumer discretionary | 55.72 | 3.31 | 64 |

Specifically, we compared the improvement on the prediction accuracy after applying late fusion. To verify the impact of news, we define the degree of improvement in prediction accuracy for news as follows:

$$\Delta\text{news} = \frac{acc_{p,n} - acc_p}{acc_p}$$

where $\text{acc}_{\text{p}}$ denotes the accuracy of multi-layer LSTM, and $\text{acc}_{\text{p,n}}$ denotes the accuracy after fusing with news content.

As shown in Fig. 6, we can see different impacts for different companies. Although not all companies have positive impacts, the accuracy is improved for most representative companies.

To verify the impact of historical stock prices, we define the degree of improvement in prediction accuracy for prices as follows:

$$\Delta\text{price} = \frac{acc_{p,n} - acc_n}{acc_n}$$

where $\text{acc}_{\text{n}}$ denotes the accuracy of HAN, and $\text{acc}_{\text{p,n}}$ denotes the accuracy after fusing with historical prices.

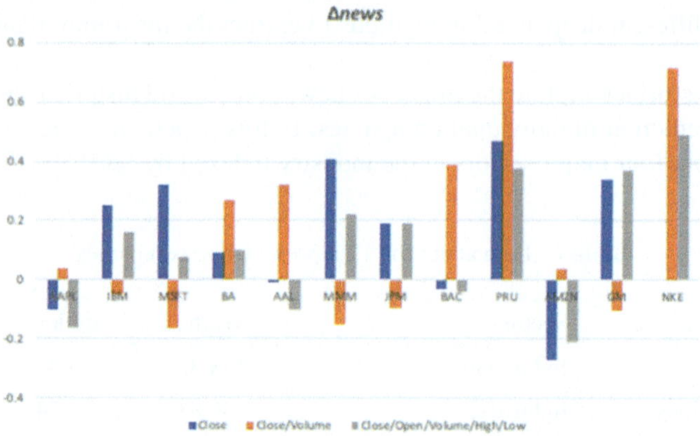


**Fig. 6.** The impact of news content on stock trend prediction

**Fig. 7.** The impact of historical prices on stock trend prediction

As shown in Fig. 7, we can see the accuracy of almost all companies is improved. The degree of improvement is higher than news content, with the highest improvement of 2.3%. This shows the potential of fusing news contents with historical prices for stock trend prediction.

## 5 Conclusions

In this paper, we proposed a deep fusion model for stock trend prediction. Firstly, we learn the semantic features from news content using multi-layer LSTM and the time-series features from historical stock prices using hybrid attention networks, respectively. Then, these two types of features are combined with early and late fusion methods. Comparing with the individual multi-layer LSTM stock prediction and news attention models, late fusion can greatly improve the performance, with the best F1-score of 79.0%. This shows the potential of our proposed method for stock trend prediction. Further investigation is needed for stock prices in different areas and countries.

## References

1. Meesad, P., Rasel, R.I.: Predicting stock market price using support vector regression. In: 2013 International Conference on Informatics, Electronics and Vision (ICIEV), pp. 1–6. IEEE (2013)
2. Kumar, I., Dogra, K., Utreja, C., Yadav, P.: A comparative study of supervised machine learning algorithms for stock market trend prediction. In: 2018 Second International Conference on Inventive Communication and Computational Technologies (ICICCT), pp. 1003–1007. IEEE (2018)
3. Hiransha, M., AbGopalakrishnan, E., Menon, V.K., Soman, K.P.: NSE stock market prediction using deep-learning models. Proc. Comput. Sci. **132**, 1351–1362 (2018)
4. Chen, K., Zhou, Y., Dai, F.: A LSTM-based method for stock returns prediction: a case study of china stock market. In: 2015 IEEE International Conference on Big Data (Big Data), pp. 2823–2824. IEEE (2015)
5. Adebiyi, A.A., Adewumi, A.O., Ayo, C.K.: Comparison of Arima and artificial neural networks models for stock price prediction. J. Appl. Math. **2014**, 7 p. (2014). https://doi.org/10.1155/2014/614342. Article ID 614342
6. Gao, T., Chai, Y., Liu, Y.: Applying long short term memory neural networks for predicting stock closing price. In: 2017 8th IEEE International Conference on Software Engineering and Service Science (ICSESS), pp. 575–578. IEEE (2017)
7. Yamini Nivetha, R., Dhaya, C.: Developing a prediction model for stock analysis. In: 2017 International Conference on Technical Advancements in Computers and Communications (ICTACC), pp. 1–3. IEEE (2017)
8. Lee, H., Surdeanu, M., MacCartney, B., Jurafsky, D.: On the importance of text analysis for stock price prediction. In: LREC, vol. 2014, pp. 1170–1175 (2014)
9. Ding, X., Zhang, Y., Liu, T., Duan, J.: Using structured events to predict stock price movement: an empirical investigation. In: Proceedings of the 2014 Conference on Empirical Methods in Natural Language Processing (EMNLP), pp. 1415–1425 (2014)
10. Hu, Z., Liu, W., Bian, J., Liu, X., Liu, T.-Y.: Listening to chaotic whispers: a deep learning framework for news-oriented stock trend prediction. In: Proceedings of the Eleventh ACM International Conference on Web Search and Data Mining, pp. 261–269 (2018)
11. Akita, R., Yoshihara, A., Matsubara, T., Uehara, K.: Deep learning for stock prediction using numerical and textual information. In: 2016 IEEE/ACIS 15th International Conference on Computer and Information Science (ICIS), pp. 1–6. IEEE (2016)
12. Shi, L., Teng, Z., Wang, L., Zhang, Y., Binder, A.: DeepClue: visual interpretation of text-based deep stock prediction. IEEE Trans. Knowl. Data Eng. **31**(6), 1094–1108 (2018)
13. Li, X., Huang, X., Deng, X., Zhu, S.: Enhancing quantitative intra-day stock return prediction by integrating both market news and stock prices information. Neurocomputing **142**, 228–238 (2014)
14. Pagolu, V.S., Reddy, K.N., Panda, G., Majhi, B.: Sentiment analysis of twitter data for predicting stock market movements. In: 2016 International Conference on Signal Processing, Communication, Power and Embedded System (SCOPES), pp. 1345–1350. IEEE (2016)
15. Sehgal, V., Song, C.: SOPS: stock prediction using web sentiment. In: Seventh IEEE International Conference on Data Mining Workshops (ICDMW 2007), pp. 21–26. IEEE (2007)
16. Zhai, Y., Hsu, A., Halgamuge, S.K.: Combining news and technical indicators in daily stock price trends prediction. In: Liu, D., Fei, S., Hou, Z., Zhang, H., Sun, C. (eds.) ISNN 2007. LNCS, vol. 4493, pp. 1087–1096. Springer, Heidelberg (2007). https://doi.org/10.1007/978-3-540-72395-0_132

17. Li, X., Wu, P., Wang, W.: Incorporating stock prices and news sentiments for stock market prediction: a case of Hong Kong. Inf. Process. Manag. **57**, 102212 (2020)
18. Deng, S., Mitsubuchi, T., Shioda, K., Shimada, T., Sakurai, A.: Combining technical analysis with sentiment analysis for stock price prediction. In: 2011 IEEE Ninth International Conference on Dependable, Autonomic and Secure Computing, pp. 800–807 (2011)
19. Xu, Y., Cohen, S.B.: Stock movement prediction from tweets and historical prices. ACL (2018)
20. Lau, J.H., Baldwin, T.: An empirical evaluation of doc2vec with practical insights into document embedding generation. arXiv preprint arXiv:1607.05368 (2016)
21. Le, Q., Mikolov, T.: Distributed representations of sentences and documents. In: International Conference on Machine Learning, pp. 1188–1196 (2014)
22. Mikolov, T., Chen, K., Corrado, G., Dean, J.: Efficient estimation of word representations in vector space. arXiv preprint arXiv:1301.3781 (2013)
23. Yang, Z., Yang, D., Dyer, C., He, X., Smola, A., Hovy, E.: Hierarchical attention networks for document classification. In: Proceedings of the 2016 Conference of the North American Chapter of the Association for Computational Linguistics: Human Language Technologies, pp. 1480–1489 (2016)

# Generating Search Text Ads from Keywords and Landing Pages via BERT2BERT

Kota Ishizuka[1(✉)], Kai Kurogi[2], Kosuke Kawakami[2], Daishi Iwai[2], and Kazuhide Nakata[1]

[1] Tokyo Institute of Technology,Tokyo, Japan
ishizuka.k.ad@m.titech.ac.jp
[2] negocia inc., Tokyo, Japan

**Abstract.** The standard way to create text ads is to capture searched keywords and the information on their landing pages (LP). However, this coupling of keywords and an LP increases the number of text ads per LP, which makes impossible to create ad texts for all effective combinations of keywords and LPs due to limitation of human resource. We propose a transformer-based ad text generation model using both keywords and LPs to reduce costs and time generating ad texts. We extract tags and texts LP's HTML, such as title, h1, h2, fine-tune a pre-trained encoder-decoder model (initialized by BERT2BERT), and HTML tag embeddings, similar to position embeddings, are passed to an input layer. The experimental results demonstrate that our model generates ad texts with a quality close to human-written ones for fluency, attractiveness, and correctness.

## 1 Introduction

Keywords and information on a landing page (LP) are the two most important references to create compelling text ads. Because search engines display the text ads linked with the keywords, the ad text not related to the keywords will lead to less attraction of the users who searched on the site. Since the users who clicked the text ads visit the landing page (LP), the ad text not associated with the LP's content will lead to fewer conversions (CVs) due to a mismatch between ad contents and the user's interest. However, it takes time and effort to create a text ad that is in line with the content of the LP and still attract the interest of users who have in-depth knowledge of the domain and products with multiple keywords.

A combination of keywords and LPs increases the number of text ads per LP. Moreover, it makes it impossible due to human resource constraints to create ad texts for all effective combinations of keywords and LPs. Hence, it often happens in practice that the same ad text is set for different keywords, or just simple keyword insertions are used. However, ad platforms take the relevance of the keyword to the ad text into account. If the relevance is high, the ad will be


Y. Takama et al. (Eds.): JSAI 2021, AISC 1423, pp. 27–33, 2022.
https://doi.org/10.1007/978-3-030-96451-1_3

displayed in a higher position, giving an advantage over competitors. It leads to a higher click-through rate (CTR). Generating ad text relevant to the keyword is one of the most critical factors in ad management.

In recent years, large pre-trained language models such as BERT [1], T5 [2], and GPT-2 [3] have been trained and published. However, most of these models have been trained in English, and available non-English models are mostly BERT pre-trained by several volunteers. Generative models, GPT-2 and T5, there are few pre-trained models in Japan. Therefore, in our work on natural language generation in Japanese, we are not using pre-trained models. In our work on natural language processing in Japanese, we are not using pre-trained models but from scratch.

We propose a transformer-based ad text generation model using both keywords and LPs to reduce costs and time generating ad texts. We design a transformer-based model that creates ad texts by referring to keywords and LPs in the same way as humans do and demonstrate possibility of replacing them with machine learning for ad text generation. Furthermore, we utilize BERT2BERT [4] to initialize a transformer-based sequence-to-sequence (seq2seq) model, which allows for a *quasi*-pre-trained encoder-decoder model only from one encoder model. The main contributions of this work are as follows:

- Applying a transformer-based model to generate ad text based on **both** keywords **and** LP.
- Proposing novel embeddings to make a model to distinguish HTML tags of an LP, like position embeddings of transformers.
- Utilizing BERT2BERT initialization to get a *quasi*-pre-trained encoder-decoder model even in a non-English language.

## 2 Related Work

The first keywords-based ad text generation system [5] uses *n-grams* and dependency trees to generate texts. In a recent study in Chinese [6], RNNs were used to generate context around a single input keyword.

For an LP-based method, *n-grams* frequently occurring together in LP are found and considered the most critical [7]. In [8], a keyword-based generation of Japanese ad texts was proposed to consider information about a product. These approaches use a way to select several words from LP, consider them as keywords, and generate ad text based on them.

The first end-to-end approach from LP to ad text was proposed in [9] with attention-based LSTM. This method does not sample keywords from LP but directly generates ad text from LP. Title and body text were extracted from LP's HTML source.

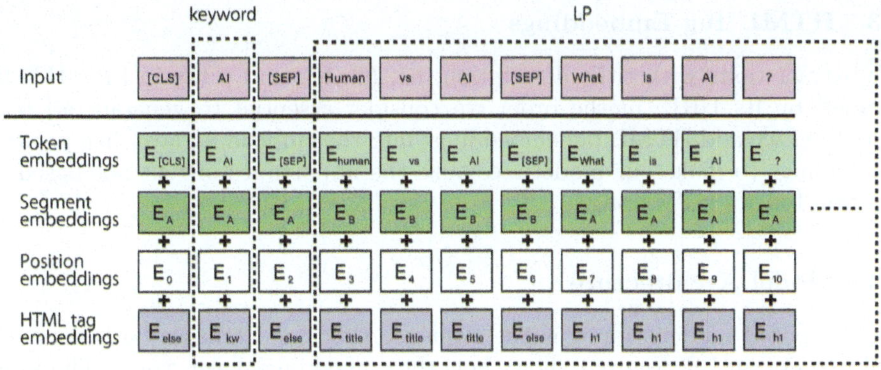


**Fig. 1.** Input layer. The input embeddings are the sum of all embeddings.

**Table 1.** Automatic and human evaluation results on the test set (see Sect. 4.2 for details). The probabilities for human evaluation results are the p-values of Mann-Whitney U test between each model and human.

| Models | Automatic evaluation | | | Human evaluation | | |
|---|---|---|---|---|---|---|
| | BLEU ↑ | ROUGE-1 ↑ | ROUGE-2 ↑ | fluency ↑ | attractiveness ↑ | correctness ↑ |
| Human | | | | 4.91 | 4.89 | 4.85 |
| LSTM | 33.5 | 39.7 | 22.8 | 3.76 ($p < .001$) | 3.89 ($p < .001$) | 3.60 ($p < .001$) |
| Transformer | 41.9 | 62.2 | 47.2 | 4.80 ($p = .241$) | 4.80 ($p = .485$) | 4.77 ($p = .475$) |

# 3 Proposed Model

## 3.1 Transformer

A transformer [10] is a neural network with permutation equivalence composed of attention mechanisms and residual blocks. In the transformer's architecture, position embeddings are generally added into the embedding layer with word embeddings since transformers cannot distinguish the input's position.

## 3.2 BERT2BERT initialization

A transformer-based seq2seq model has two transformers, an encoder, and a decoder. The difference between an encoder and a decoder is that only decoders have a source-target-attention (STA) layer. Therefore, BERT checkpoints have weights of all layers except STA layers.

BERT2BERT [4] initializes the encoder with BERT checkpoint's weights and all layers except STA layers with the same weights as the encoder. However, STA layers are randomly initialized. We use the checkpoint of [11]'s BERT base pre-trained model.

### 3.3 HTML Tag Embeddings

We extract texts with title, description, h1, h2, h3, and p tag in LP's HTML, scraped by its URL, in the order we consider essential to generate ad text. Moreover, we add HTML tag embeddings into the embedding layer like position embeddings to make our model recognize the importance of text not based on position but its HTML tag.

### 3.4 Model Architecture

The proposed model's input is the sequence that concatenates both keywords and texts of an LP. Figure 1 shows how the embeddings layer works. The input embeddings are the sum of token embeddings, segmentation embeddings, position embeddings, and HTML tag embeddings.

## 4 Experiments

### 4.1 Setting

For training and test data, we used data obtained by our scraping process. Each data consists of a searched query (keywords), title and description displayed as a listing advertisement, and HTML of the LP. The number of the training data is 741,506, and that of the test is 82,320. These data are collected by scraping methods with keywords in our internal search query list.

We compared a baseline and the proposed model. The baseline is our implementation of [9] without reinforcement learning (RL) since the RL model generates repeated tokens. Plus, we conducted an ablation study to demonstrate the effectiveness of our proposed methods, input keywords, input LP, HTML tag embeddings, and BERT2BERT initialization.

**Table 2.** Ablation study results on the test set.

| Elements of the proposed methods | | | | Automatic evaluation | | |
|---|---|---|---|---|---|---|
| input keywords | input LP | HTML tag embeddings | BERT2BERT initialization | BLEU ↑ | ROUGE-1 ↑ | ROUGE-2 ↑ |
| ✓ | ✓ | ✓ | ✓ | 41.9 | 62.2 | 47.2 |
| ✓ | ✓ | ✓ | ✗ | 31.0 | 54.3 | 36.3 |
| ✓ | ✓ | ✗ | ✓ | 39.8 | 60.4 | 44.9 |
| ✓ | ✗ | ✓ | ✓ | 6.78 | 27.9 | 10.1 |
| ✗ | ✓ | ✓ | ✓ | 32.9 | 55.9 | 40.3 |

**Table 3.** Examples of generated ad texts in Japanese/English.

| | | |
|---|---|---|
| Keywords | | Amazon 広告運用 AI |
| LP's url | | https://commerce-flow.com |
| **Keywords** | **LP** | **Generated ad text** |
| ✓ | ✗ | 【公式】Amazon 広告運用 ON - [NUM]ヶ月お試しプログラム有<br>相談やお見積りは無料ですので、まずはご連絡くださいませ。<br>[Official] Amazon Ad Management ON - [NUM] Month Trial Program Available<br>Consultations and estimates are free, so please contact us first. |
| ✗ | ✓ | 【国産】スポンサー広告なら - 業界最高レベルの自動化が可能<br>専門知識は不要で、数ステップの設定や完了すれば最短即日初期費用/[NUM] 円。<br>[Domestic] For Sponsored Ads - The highest level of automation in the industry<br>No technical knowledge required, just a few steps to set up and complete,<br>initial cost/[NUM] yen in the shortest possible time. |
| ✓ | ✓ | 【国産 No.[NUM]】スポンサー広告 - 最先端 AI でクリック率アップ<br>専門知識不要の最先端 AI エンジンによる自動入札・<br>自動解析で競合他社との圧倒的な差別化に。<br>Japan's No. [NUM] Sponsored Ads - Increase CTRs with state-of-the-art AI<br>Automatic bidding and analysis by the most advanced AI engine<br>that requires no technical knowledge. Unparalleled differentiation from competitors |

### 4.2 Automatic and Human Evaluations

Table 1 shows the results of automatic and human evaluations. We first evaluate the models with popular natural language generation evaluation metrics such as BLEU [12], and ROUGE [13]. Additionally, we conducted human evaluations with four workers in negocia, Inc. They scored from one to five points in fluency, attractiveness, and correctness for forty sentences.

Our transformer-based model outperformed the baseline in both automatic and human evaluation. Furthermore, each null hypothesis that the median of the proposed model's human evaluation scores is the same as that of humans is not rejected at a 5% significance level. It implies that our model's quality of ad text is as good as that of humans.

### 4.3 Ablation Study

Through an ablation study, we show the effectiveness of our proposed method and other effective techniques. Table 2 demonstrates that the most significant contribution is to put both keywords and texts of LP into the model. The second is BERT2BERT, and the third is HTML tag embeddings. All of them contribute to the improvement of the automatic evaluation metrics.

Table 3 shows an example of ad text with abbreviations. There are few grammatical problems with the keyword-only model, but when the LP is ignored, the ad text does not include the product description. On the other hand, in the model using only LPs, the product description is included, but users' interest in "AI for ad management" is missing. It also looks like just a service that performs ad automation. The model that uses both keywords and LP effectively combines a user's interest and explains the product content well.

## 5 Conclusion

We presented a transformer-based model to generate ad text from both keywords and an LP. With the help of BERT2BERT initialization and HTML tag embeddings, our model aptly generates text ads more fluent, attractive, and correctly than the LSTM model trained from scratch and much closer to humans. We showed the importance of both keywords and an LP to put into the model through our evaluation. As shown in many other domains, we confirmed that a transformer-based model is better than LSTM. Its pre-trained checkpoints are available even for non-English languages and can be converted to a *quasi*-pre-trained seq2seq model with BERT2BERT initialization. Future work could include using this method in conjunction with reinforcement learning to outperform humans.

## References

1. Devlin, J., Chang, M.-W., Lee, K., Toutanova, K.: Bert: Pre-training of deep bidirectional transformers for language understanding. arXiv preprint arXiv:1810.04805 (2018)
2. Raffel, C., et al.: Exploring the limits of transfer learning with a unified text-to-text transformer. arXiv preprint arXiv:1910.10683 (2019)
3. Radford, A., Jeffrey, W., Child, R., Luan, D., Amodei, D., Sutskever, I.: Language models are unsupervised multitask learners. OpenAI blog **1**(8), 9 (2019)
4. Rothe, S., Narayan, S., Severyn, A.: Leveraging pre-trained checkpoints for sequence generation tasks. Trans. Assoc. Comput. Linguist. **8**, 264–280 (2020)
5. Uchimoto, K., Sekine, S., Isahara, H.: Text generation from keywords. In: COLING 2002: The 19th International Conference on Computational Linguistics (2002)
6. Song, Z., Liu, L., Song, W., Zhao, X., Du, C.: A neural network model for Chinese sentence generation with key word. In: 2019 IEEE 9th International Conference on Electronics Information and Emergency Communication (ICEIEC), pp. 334–337 (2019)
7. Thomaidou, S., Lourentzou, I., Katsivelis-Perakis, Vazirgiannis, M.: Automated snippet generation for online advertising. In: Proceedings of the 22nd ACM international conference on Information and Knowledge Management, pp. 1841–1844 (2013)
8. Wakimoto, K., Kawamoto, S., Zhang, P.: Keyword-based text generation for internet advertisement. In: The 34th Annual Conference of the Japanese Society for Artificial Intelligence, 2020, JSAI2020:4Rin170–4Rin170 (2020)
9. Hughes, J.W., Chang, K., Zhang, R.: Generating better search engine text advertisements with deep reinforcement learning. In: Proceedings of the 25th ACM SIGKDD International Conference on Knowledge Discovery and Data Mining, KDD 2019, pp. 2269-2277. Association for Computing Machinery, New York, NY, USA (2019)

10. Vaswani, A., et al.: Attention is all you need. arXiv preprint arXiv:1706.03762 (2017)
11. Zhao, X., Hamamoto, M., Fujihara, H.: Laboro Bert Japanese: Japanese Bert pre-trained with web-corpus. https://github.com/laboroai/Laboro-BERT-Japanese (2020)
12. Papineni, K., Roukos, S., Ward, T., Zhu, W.-J.: Bleu: a method for automatic evaluation of machine translation. In: Proceedings of the 40th Annual Meeting of the Association for Computational Linguistics, pp. 311–318 (2002)
13. Lin, C.-Y.: Rouge: a package for automatic evaluation of summaries. In: Text Summarization Branches Out, pp. 74–81 (2004)

# Machine Translation Utilizing Similar Translation Retrieval

Takuya Tamura[1], Yizhen Wei[1], Takehito Utsuro[1(✉)], and Masaaki Nagata[2]

[1] Graduate School of Science and Technology, University of Tsukuba, 1-1-1, Tennodai, Tsukuba, Ibaraki 305-8573, Japan
utsuro@iit.tsukuba.ac.jp

[2] NTT Communication Science Laboratories, NTT Corporation, 2-4, Hikaridai, Seika-cho, "Keihanna Science City", Kyoto 619-0237, Japan

**Abstract.** Neural Fuzzy Repair (NFR) system enables an NMT system to improve translation accuracy using similar translations searched from Translation Memory. This paper compared edit distance and sentence-BERT (SBERT) as the similarity measures used in the search for similar translations, and showed that SBERT outperformed edit distance in the case of small corpus sizes. This paper also studied a method to automatically select the most appropriate translation from more than one candidates. Compared to the naive method based on the number of tokens, the method based on the inner product of SBERT's sentence embedding achieved significant improvements. These results prove the effectiveness of the SBERT-based approach in the NFR system.

## 1 Introduction

Translation Memory (TM) is a set of high quality manually translated sentence pairs of source and target language sentences, and is used in a variety of situations. In the past, TM has been used in Computer-Aided Translation (CAT) tools to assist human translators. Using a translation memory, sentences stored in TM can be translated without errors by simply replacing it with the translated sentence. In addition, because of its usefulness in translating similar sentences, the incorporation of TM into Neural Machine Translation (NMT) [1,11] has been studied in recent years.

In this paper, we evaluate the Neural Fuzzy Repair (NFR) model of [1] which incorporates TM into NMT, and evaluate retrieval methods of similar translations. We also propose methods to generate multiple translation candidates by using multiple similar sentences, and to automatically select the most appropriate translation from them. In the evaluation of English to French translation using the ECB corpus dataset [8][1], when the size of the training set was small, sentence-BERT (SBERT) based methods achieved significant improvements in the evaluation of searching similar sentences from the TM and selecting the most appropriate translation.

[1] http://opus.nlpl.eu/ECB.php.


Y. Takama et al. (Eds.): JSAI 2021, AISC 1423, pp. 34–44, 2022.
https://doi.org/10.1007/978-3-030-96451-1_4

## 2 Related Work

Bulte and Tezcan [1] proposed an NFR model that improves translation accuracy by incorporating TM into NMT. This model retrieves a similar sentence based on edit distance from the set of source language sentences in the TM, and concatenates the source sentence with the target language translation of the similar sentence as an input to NMT. When searching for similar sentences, the lower bound of the similarity score is selected so that similar sentences can be found in half of the source language sentences of the training dataset. This indicates that the NFR model is trained only on cases where similar sentences with a relatively high similarity score are available. Therefore, if there is no similar sentence found in the decoding phase, the baseline NMT model without NFR is used as a back-off. In this model, TM can be incorporated into NMT without modifying the NMT model architecture itself, simply by prepossessing the data input to the NMT model. Therefore, it is highly compatible with any existing NMT model and has high portability in terms of implementation.

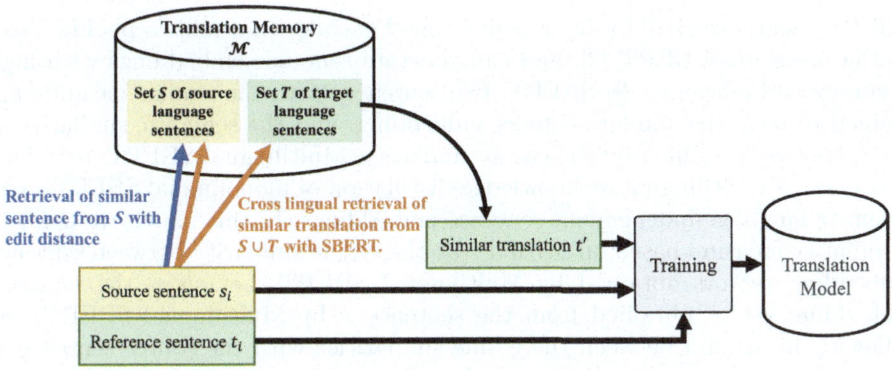


**Fig. 1.** Framework of translation model training by NFR

## 3 Similarity Measures of Similar Translations

In this section, we describe similarity measures for retrieving similar sentences from a TM. Let a TM $\mathcal{M} = \{(s_i, t_i)|i = 1, \ldots, n\}$ be a set of sentence pairs of source sentence $s_i$ and target sentence $t_i$. Let $S = \{s_i|i = 1, \ldots, n\}$ be the set of source language sentences, and $T = \{t_i|i = 1, \ldots, n\}$ be the set of target language sentences. The similarity measures are defined as a mapping from a source sentence $s$ and a sentence pair $(s', t')$ in the TM to a real value representing the similarity.

$$sim : S \times \mathcal{M} \rightarrow \mathbb{R}$$

In this section, we define edit distance based and SBERT based similarity measures, respectively.

### 3.1 Similarity Measure based on Edit Distance

The edit distance is defined as the minimum number of operations required to convert the original string into another string by inserting, deleting, or replacing a single character. In this paper, we followed [1] and adopted the following similarity score of [9]

$$sim(s,(s',t')) = 1 - \frac{\Delta_{ed}(s,s')}{\max(|s|_c, |s'|_c)}$$

where $\Delta_{ed}(s,s')$ is the edit distance between two sentences $s, s'$, and $|s|_c$ is the number of characters in $s$. When $s$ and $s'$ perfectly match, the similarity score takes the maximum value $sim(s,s') = 1$. Since the edit distance can only be calculated between two sentences of the same language, the search range is limited to the set $S$ of source sentences in the TM.

### 3.2 Cosine Similarity Using Sentence-BERT

SBERT (Sentence-BERT) [5][2] is a fine-tuned encoder that adds a pooling layer to the pre-trained BERT [2], and can generate sentence embeddings with high accuracy and efficiency. In SBERT, two sentences with the similar meaning are trained to have the similar sentence embedding, and the semantic similarity of the sentences is defined by the cosine similarity. Multilingual SBERT [6], which is trained as multilingual by knowledge distillation of monolingual SBERTs, can generate language-independent sentence embeddings. In this paper, we define a "similarity measure based on SBERT" by the cosine similarity between sentence embedding vectors obtained by Multilingual SBERT. Let $sb(x)$ the sentence embedding vector obtained from the sentence $x$ by Multilingual SBERT, the following maximum between the cosine similarities with the source sentence $s'$ and with the target sentence $t'$ is used:

$$sim(s,(s',t')) = \max_{v \in \{s',t'\}} \cos(sb(s), sb(v))$$

In Sect. 5.3, we use the cosine similarity by Multilingual SBERT in automatic selection of the most appropriate translation from multiple translation candidates.

## 4 Neural Fuzzy Repair (NFR) [1]

Following the framework (Fig. 1) of NFR (Neural Fuzzy Repair) by [1], a similar target sentence retrieved based on the similarity measure $sim$ from the TM $\mathcal{M}$, then concatenated with the source sentence and is finally input to NMT. Thus, the TM $\mathcal{M}$ is used both as the database for retrieving similar translations and as the dataset for training the translation model.

[2] https://github.com/UKPLab/sentence-transformers.

### 4.1 Training

Let $s_i(i = 1, \ldots, n)$ be the source sentence, $t_i(i = 1, \ldots, n)$ be its translation, and $(s_i, t_i)(\in \mathcal{M})$ be the translation pair of $s_i$ and $t_i$. First, the sentence $t'$ in the target language that is similar to the source sentence $s_i$ $(\exists s', (s', t') \in \mathcal{M})$ is retrieved. Then, the similar sentence pair $(s', t')$ that maximizes the similarity measure $sim$ satisfying the similarity lower bound $lbd$ $(sim(s_i, (s', t')) \geq lbd)$ is retrieved.[3] When $(s', t')$ is retrieved, following [1], $t'$ is concatenated with $s_i$ as below:

$$s_i@@@t'$$

Finally, the translation model $m_{nfr}$ of NFR is trained with $(s_i@@@t', t_i)$ $(i = 1, \ldots, n)$.

### 4.2 Decoding

The decoding procedure is represented as the following formula, where the source sentence is translated by the translation model $m_{nfr}$ of NFR if any similar translation pair is found, and otherwise translated by the baseline translation model $m_b$:

$$\text{NFR}(s, m_b, m_{nfr}, lbd) = \begin{cases} trans(m_{nfr}, s\ @@@\ t') \\ \qquad\qquad (sim(s, (s', t')) \geq lbd) \\ trans(m_b, s) \\ \qquad\qquad (sim(s, (s', t')) < lbd) \end{cases}$$

## 5 Methods of Automatic Selection of the Most Appropriate Translation

In NFR, multiple translation candidates can be obtained by utilizing multiple similar translations. This section introduces methods of automatic selection of the most appropriate translation $\hat{u}$ from the set $U = \{u_i | i = 1, \ldots, m\}$ of candidate translations.

### 5.1 Translation with Maximum Number of Tokens

Let $|u|_w$ be the number of tokens of the translation $u$, that with the maximum number of tokens is selected as below:

$$\hat{u} = \underset{u \in U}{\text{argmax}}\, |u|_w$$

[3] Except for the case of the perfect match with $t_i$ $(t' \neq t_i)$. In the case of the edit distance as the similarity measure, we follow the condition of $sim(s_i, (s', t')) > lbd = 0$ when the lower bound as $lbd = 0$.

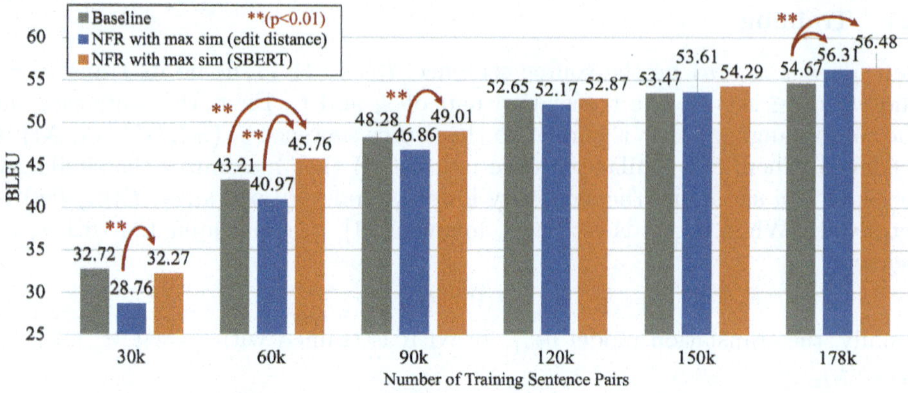


**Fig. 2.** Evaluation results of NFR models

### 5.2 Translation Closest to the Estimated Number of Tokens

Let $s$ be the source sentence, $k$ be the ratio of the numbers of tokens of the source and target sentences within the training sentence pairs, translation that is closest to the estimated number of tokens is selected as below:[4]

$$\hat{u} = \underset{u \in U}{\operatorname{argmin}} \Big| |u|_w - k|s|_w \Big|$$

$$k = \frac{\sum_{t \in T} |t|_w}{\sum_{s \in S} |s|_w}$$

### 5.3 Maximizing SBERT Cosine Similarity with the Source Sentence

In this approach, translation that maximizes the SBERT cosine similarity with the source sentence $s$ is selected as below:

$$\hat{u} = \underset{u \in U}{\operatorname{argmax}} \cos(sb(s), sb(u))$$

## 6 Evaluation

### 6.1 Evaluation Procedure

In the evaluation, English to French version of ECB corpus [8] (See footnote 1) is used. The numbers of the training sentence pairs are, 30k, 60k, 90k, 120, 150k,

[4] In the evaluation of this paper, the ratio $k = 1.18$ measured from the ECB corpus [8] is used.

**Table 1.** Outline of the Evaluation Results of the Methods of Automatic Selection of the Most Appropriate Translation (lower bound of similarity = 0, ○: significantly outperforming the baseline and NFR ($p < 0.01$, or $p < 0.05$), ×: other than ○)

| Methods of automatic selection of the most appropriate translation | Similarity measures | Number of training sentence pairs | | | | | |
|---|---|---|---|---|---|---|---|
| | | 30k | 60k | 90k | 120k | 150k | 178k |
| Translation with maximum number of tokens | Edit distance | × | × | × | × | × | × |
| | SBERT | ○ | × | × | × | × | × |
| | Edit distance + SBERT | ○ | × | × | × | × | × |
| Translation closest to the estimated number of tokens | Edit distance | × | × | × | × | × | × |
| | SBERT | ○ | × | × | × | × | × |
| | Edit distance + SBERT | ○ | × | × | × | × | × |
| Maximizing SBERT cosine similarity with the source sentence | Edit distance | ○ | × | × | × | × | × |
| | SBERT | ○ | ○ | ○ | ○ | × | × |
| | Edit distance + SBERT | ○ | × | × | ○ | × | × |
| Oracle | Edit distance | ○ | × | ○ | ○ | ○ | ○ |
| | SBERT | ○ | ○ | ○ | ○ | ○ | ○ |
| | Edit distance + SBERT | ○ | ○ | ○ | ○ | ○ | ○ |

and 178k, while those of development and evaluation sentences are 1,000 each. In the training, the similar translation with the maximum similarity score is used, while in the decoding, the similar translations with the maximum, the second maximum, and the third maximum similarity scores are used.

OpenNMT-py[5] is used as the NMT platform, while Transformer [10] is used as the baseline NMT model, where the number of head is as 4, 6 layers each for encoder and decoder, the embedding size as 512 and the hidden size as 2,048, For training, we used the Adam optimizer [3] with a learning rate of 1, warm-up of 4,000, and the dropout of 0.3. Each batch contains up to 4,096 tokens. The training is stopped when the loss of the development sentences does not decrease for 10 steps. Nvidia Geforce TITAN RTX 24GB GPU is used as the hardware. Both English and French sentences are tokenized by Moses Tokenizer [4][6] and then segmented into subwords by BPE (Byte Pair Encoding) [7][7].

In the evaluation of the methods of automatic selection of the most appropriate translation from multiple translation candidates, the following 9 cases are evaluated:

(i) The translation by NFR with the maximum similarity based on edit distance and that by the baseline.
(ii) The translation by NFR with the maximum similarity based on SBERT and that by the baseline.
(iii) Three translations of (i) and (ii) in total.

[5] https://github.com/OpenNMT/OpenNMT-py.
[6] https://github.com/moses-smt/mosesdecoder.
[7] https://github.com/rsennrich/subword-nmt.

(iv) The translations by NFR with the maximum and the second maximum similarities based on edit distance and that by the baseline.
(v) The translations by NFR with the maximum and the second maximum similarities based on SBERT and that by the baseline.
(vi) Five translations of (iv) and (v) in total.
(vii) The translations by NFR with the maximum, the second maximum, and the third maximum similarities based on edit distance and that by the baseline.
(viii) The translations by NFR with the maximum, the second maximum, and the third maximum similarities based on SBERT and that by the baseline.
(ix) Seven translations of (vii) and (viii) in total.

The translation with the maximum Sentence-BLEU score against the reference translation is selected as the oracle translation. BLEU score evaluation and the statistical significance test are done with the script of the Moses decoder[8] and the MTEval Toolkit[9], respectively.

### 6.2 Results

Figure 2 shows the overall evaluation results. In the ranges with relatively small number of training sentence pairs, NFR with SBERT based similarity significantly ($p < 0.01$) outperforms that with edit distance based similarity. Against the baseline, edit distance based similarity achieved significant ($p < 0.01$) improvement only with the number of training sentence pairs as 178k, while SBERT based similarity achieved significant ($p < 0.01$) improvement with the numbers of training sentence pairs as 60k and 178k.

Table 1 shows the overall evaluation results of the methods of automatic selection of the most appropriate translation from multiple translation candidates. It is impressive to note that "maximizing SBERT cosine similarity with the source sentence" achieved significant improvement with the numbers of training sentence pairs as 30k–120k. Figures 3, 4, 5 and 6 show the detailed evaluation results of each case. In particular, for the case with 30k training sentences in Fig. 3, BLEU improved by up to 2.28 points compared to the baseline.

Oracle, on the other hand, achieved the significant improvements with the numbers of training sentence pairs other than 60k. Figures 7, 8 and 9 show the detailed evaluation results of the method of "Oracle" with the number of training sentence pairs as 30k, 120k and 178k. In the case with 30k shown in Fig. 7, BLEU improved by up to 6.43 points compared to the baseline and by up to 6.88 points compared to the NFR (SBERT) with the maximum similarity. In the case with 120k shown in Fig. 8, BLEU improved by up to 7.12 points compared to the baseline and by up to 6.90 points compared to the NFR (SBERT) with the maximum similarity. In the case with 178k shown in Fig. 9, BLEU improved by up to 8.78 points compared to the baseline and by up to 6.97 points compared to the

[8] multi-bleu.perl.
[9] https://github.com/odashi/mteval.

NFR (SBERT) with the maximum similarity. Therefore, these results indicate that further improvement is expected for the methods of automatic selection of the most appropriate translation from multiple translation candidates.

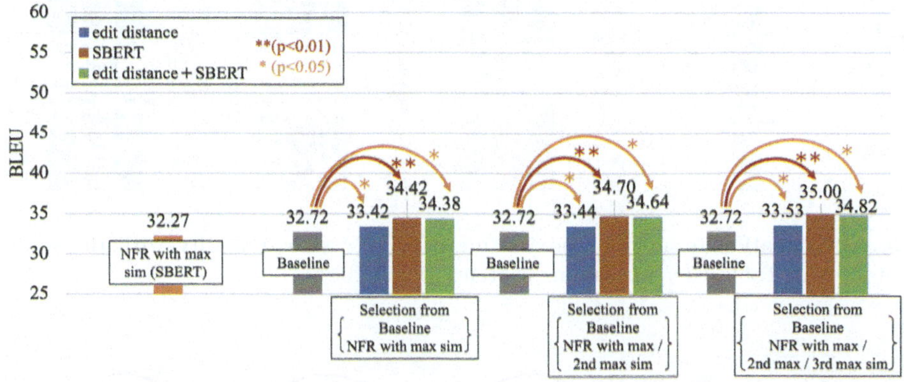


**Fig. 3.** Evaluation results of "maximizing SBERT cosine similarity with the source sentence" in Table 1 (# of training sentence pairs: 30k)

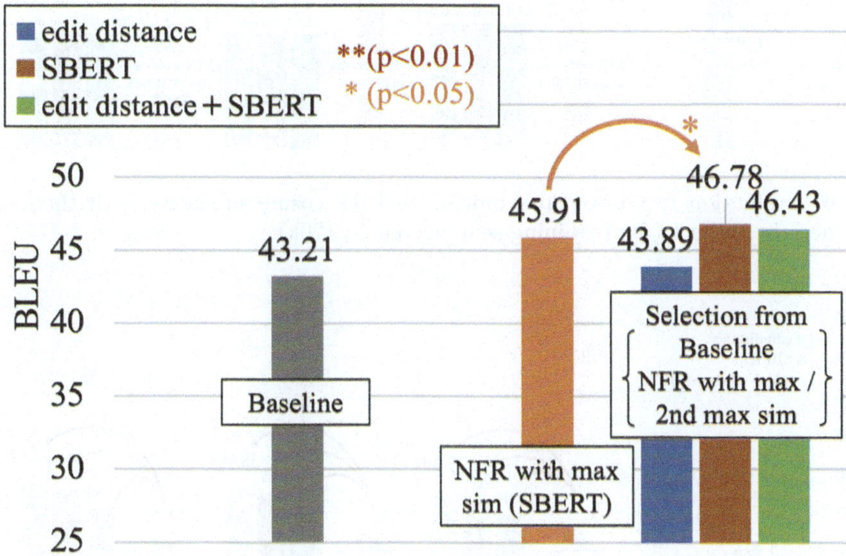


**Fig. 4.** Evaluation results of "maximizing SBERT cosine similarity with the source sentence" in Table 1 (# of training sentence pairs: 60k)

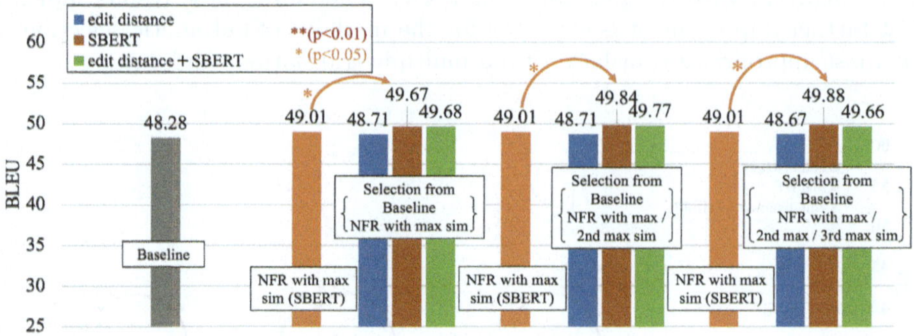


**Fig. 5.** Evaluation results of "maximizing SBERT cosine similarity with the source sentence" in Table 1 (# of training sentence pairs: 90k)

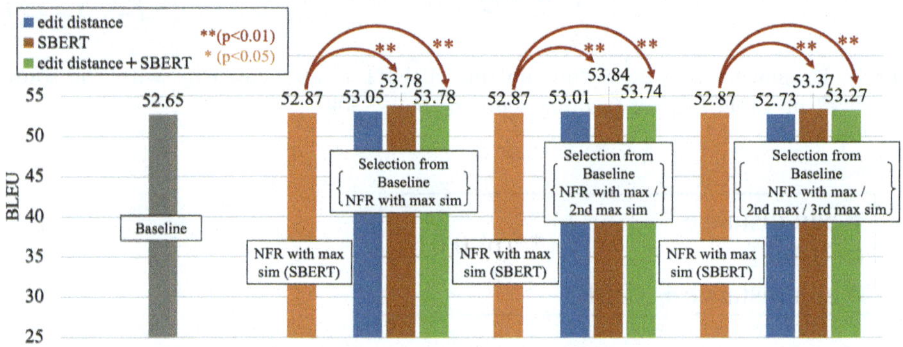


**Fig. 6.** Evaluation results of "maximizing SBERT cosine similarity with the source sentence" in Table 1 (# of training sentence pairs: 120k)

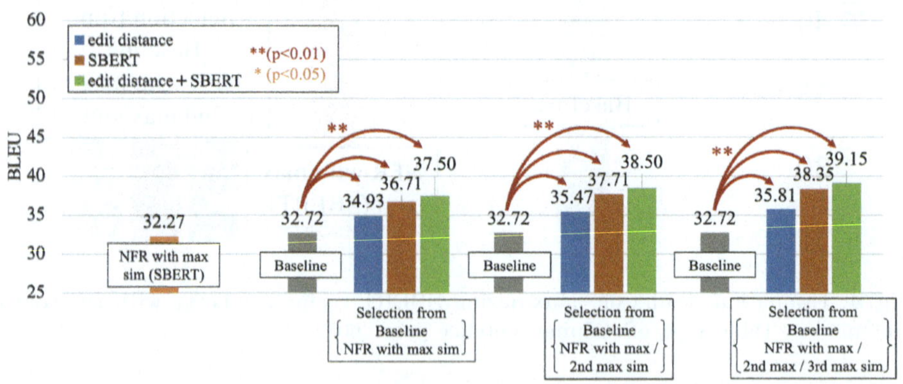


**Fig. 7.** Evaluation results of "Oracle" in Table 1 (# of training sentence pairs: 30k)

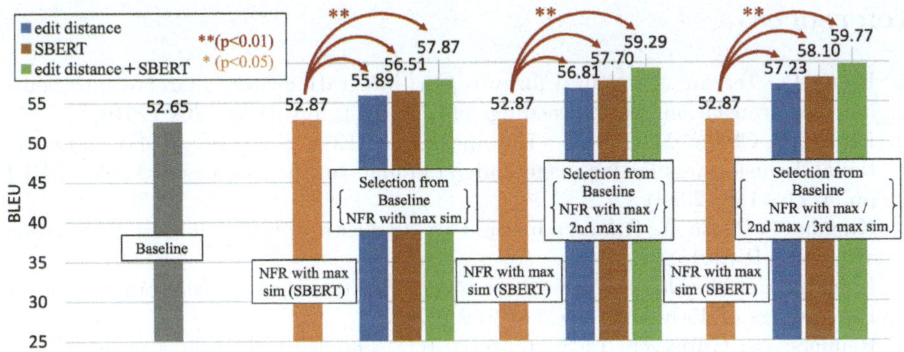


**Fig. 8.** Evaluation results of "Oracle" in Table 1 (# of training sentence pairs: 120k)

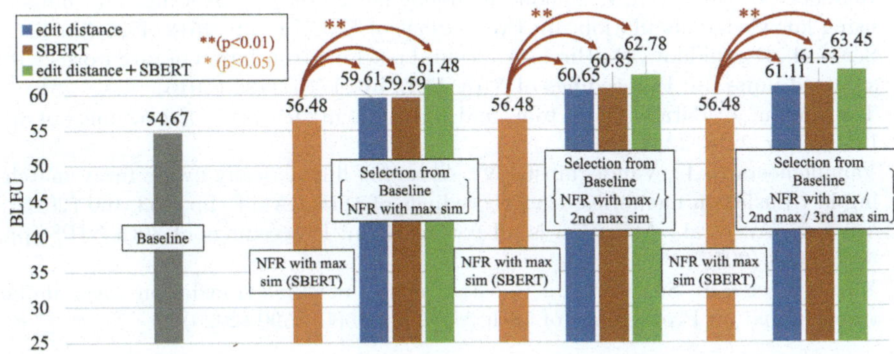


**Fig. 9.** Evaluation results of "Oracle" in Table 1 (# of training sentence pairs: 178k)

## 7 Conclusion

Neural Fuzzy Repair (NFR) system enables an NMT system to improve translation accuracy using similar translations searched from Translation Memory. This paper compared edit distance and SBERT as the similarity measures used in the search for similar translations, and showed that SBERT outperformed edit distance in the case of small corpus sizes. This paper also studied a method to automatically select the most appropriate translation from more than one candidates. Compared to the naive method based on the number of tokens, the method based on the inner product of SBERT's sentence embedding achieved significant improvements. These results proved the effectiveness of the SBERT-based approach in the NFR system.

## References


1. Bulte, B., Tezcan, A.: Neural fuzzy repair: integrating fuzzy matches into neural machine translation. In: Proceedings of 57th ACL, pp. 1800–1809 (2019)
2. Devlin, J., Chang, M., Lee, K., Toutanova, K.: BERT: pre-training of deep bidirectional Transformers for language understanding. In: Proceedings of NAACL-HLT, pp. 4171–4186 (2019)
3. Kingma, D.P., Ba, J.: Adam: a method for stochastic optimization. In: Proceedings of 3rd ICLR (2015)
4. Koehn, P., et al.: Moses: open source toolkit for statistical machine translation. In: Proceedings of 45th ACL, pp. 177–180 (2007)
5. Reimers, N., Gurevych, I.: Sentence-BERT: sentence embeddings using Siamese BERT-networks. In: Proceedings of EMNLP and 9th IJCNLP, pp. 3982–3992 (2019)
6. Reimers, N., Gurevych, I.: Making monolingual sentence embeddings multilingual using knowledge distillation. In: Proceedings of EMNLP, pp. 4512–4525 (2020)
7. Sennrich, R., Haddow, B., Birch, A.: Neural machine translation of rare words with subword units. In: Proceedings of 54th ACL, pp. 1715–1725 (2016)
8. Tiedemann, J.: Parallel data, tools and interfaces in OPUS. In: Proceedings of 8th LREC, pp. 2214–2218 (2012)
9. Vanallemeersch, T., Vandeghinste, V.: Assessing linguistically aware fuzzy matching in translation memories. In: Proceedings of 18th EAMT, pp. 153–160 (2015)
10. Vaswani, A., et al.: Attention is all you need. In: Proceedings of 30th NIPS, pp. 5998–6008 (2017)
11. Xu, J., Crego, J., Senellart, J.: Boosting neural machine translation with similar translations. In: Proceedings of 58th ACL, pp. 1580–1590 (2020)

# Automatic Utterance Selection Based on Prosodic Features of Children's Vocalizations

Yuki Kubo, Natsuki Oka(✉), Subaru Hanada, Kazuaki Tanaka, and Tomomi Takahashi

Kyoto Institute of Technology, Matsugasaki, Sakyo-ku, Kyoto 606-8585, Japan
y-kubo@ei.sanken.osaka-u.ac.jp, nat@kit.ac.jp

**Abstract.** There is a growing need for dialogue systems that can interact appropriately with children. In some dialogue systems, the system selects an utterance in a given scenario. Our goal is to use prosody to generate an appropriate dialogue without requiring a user's dialogue selection. In this study, we constructed a learning system for an utterance selection in a scenario called "Introduction". The developed system learns the dialogue selections from the prosody of a the mother's selection of dialogues and a child's responses. In our experiments, we found that by using prosodic information for the state, the system can learn new dialogue selections.

**Keywords:** Dialogue system · Reinforcement learning · Transfer learning · Unsupervised learning

## 1 Introduction

In recent years, a variety of systems capable of communicating with humans have been developed and the demand for systems that can perform spoken dialogue has increased. Additionally, the need for systems that can interact with children is even greater based on the potential decrease in child interactions caused by infectious disease outbreaks. Using a voice dialogue system to interact with children on behalf of their caregivers would prevent children from becoming isolated in scenarios where it is difficult for them to have direct interactions. Therefore, there is a growing need for dialogue systems that can appropriately interact with children.

A system that automatically interacts with the child plays the role of talking with the child on behalf of their parent. One of the goals of constructing systems that interact with children is to reduce the burden of childcare. Our goal is to select dialogues automatically. In this study, we attempted to use children's reactions to improve speech choices.

The system that was developed for automatic dialogue with children in our previous study [1] does not use the prosodic information regarding a child's


Y. Takama et al. (Eds.): JSAI 2021, AISC 1423, pp. 45–56, 2022.
https://doi.org/10.1007/978-3-030-96451-1_5

speech, so it does not necessarily learn optimally. The inputs for our previous system [1] include internal microphone inputs and histories of dialogue selections by caregivers. Additionally, in an automatic spoken dialogue system for children, it would be desirable to use speech inputs that are less affected by user positions. Therefore, the goal of this study was to construct an automatic dialogue instruction system that uses prosodic information to learn appropriate speech responses by learning according to the prosodic information of a child's responses. To realize this goal, we also aimed to evaluate the expected value of rewards as a result of reinforcement learning using prosody by making comparisons under multiple conditions in prosody-based and non-prosody-based cases.

## 2 Related Work

For dialogue systems that interact with users, various approaches have been adopted to create systems that provide more appropriate responses. In particular, the creation of systems that respond more appropriately by using information on emotions, and methods for obtaining information, such as emotional information, have been studied extensively. An emotional chatting machine [2] provides accurate responses by considering user emotions. MoEL [3] can perform end-to-end emotion-based dialogue. There is also a system that performs emotion classification by tracking individual parties [4]. There are studies that have estimated user interest in a topic using multimodal information [5,6]. There is a system that uses unsupervised learning [7] to estimates human emotions and imitates emotional expressions. Our study is unique in that we are constructing a dialogue system for children. When interacting with children, it is important to consider that they may not always respond to the system. It is also important to consider that children may move around during interaction. Prosodic information is less affected by a child's position. By focusing on prosodic information, we aim to learn spoken dialogue from children's responses.

There are several methods available for transfer learning, and transfer learning is also used in neural networks [8,9]. Transfer learning in neural networks is not only used in the image field, but also in the speech field [10]. In this study, we used transfer learning to reduce the impact of inaccurate labels. Supervised learning with transfer learning was used to assign rewards. In our transfer learning method, our neural network was first trained on labeled data from the Hazumi dataset[1], which served as the source domain. Next, the trained neural network was retrained on labeled data from the target domain and children's interactions. Clustering using the k-means method was performed to assign states.

## 3 Method

### 3.1 Whole System

In this paper, we propose a method that uses additional prosodic information on children's responses to extend the method from a previous study [1]. An overview

[1] https://github.com/ouktlab/Hazumi1712/.

of the reinforcement learning method for automatic dialogue is presented in Fig. 1. In this study, we focused on the expected value of the reward at which reinforcement learning converges. If the first response to a previous machine utterance is a child utterance, then prosody is used. Otherwise, prosody is not used. In our system, we assume an immediate reward with a discount rate of $\gamma = 0$. The distribution of utterance selection is obtained using the following equation, where $R(a_k, s_j)$ and $R(a_i, s_j)$ is the reward expectation.

$$P(a_i, s_j) = \frac{\exp\left(\frac{R(a_i, s_j)}{\tau}\right)}{\sum_{k \in A} \exp\left(\frac{R(a_k, s_j)}{\tau}\right)} \tag{1}$$

The results were checked at $\tau = 0.01$, 0.1, 1, and 10. We selected the value of $\tau$ = 1, which yieled results closest to a mother's utterance selection.

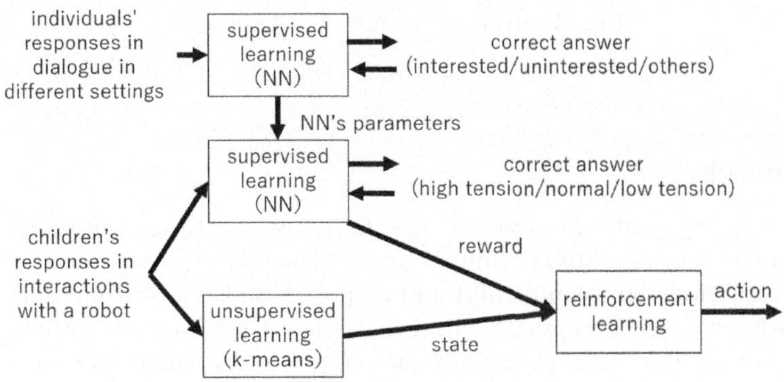


**Fig. 1.** Whole system.

### 3.2 Assigning Rewards and States

In the proposed system, prosodic information is used to assign rewards and states. For reward assignment, we used a neural network with transfer learning. The structure of the neural network is illustrated in Fig. 2. The source domain for transfer learning is the Hazumi dataset, and the target domain is data on mother-child interactions using a machine. The following formula was used for reward assignment.

$$r = \frac{0.5a + b + 1.5c}{a + b + c} \tag{2}$$

In this equation, a, b, and c correspond to the values of the neurons corresponding to a: sad or boring prosody, b: normal tension prosody, and c: happy or joyful prosody, respectively.

The k-means method, labels, and neural network are used for assigning states. For the k-means method, we only use data on mother-child interactions through a machine as training data.

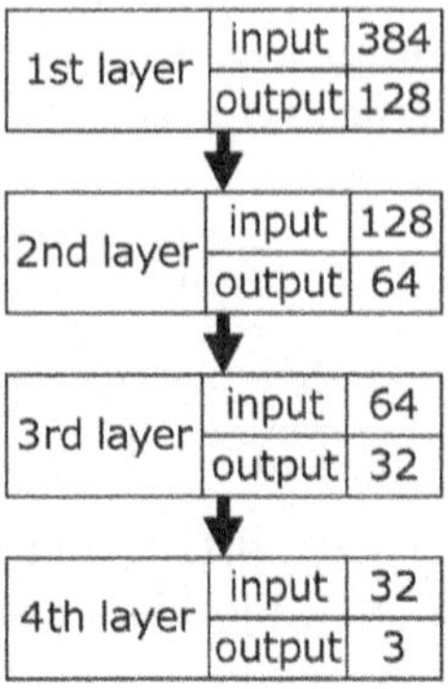


**Fig. 2.** Structure of the neural network.

### 3.3 Dataset

In this study, we used two datasets. The first was a dataset containing interactions between a system and a child [11,12]. These data are unique in that they record interactions between a child and a spoken dialogue system. This dataset was recorded because parents do not always select the optimal utterance choices.

To generate this dataset, a mother's choice of machine utterances and a child's reactions were recorded. For dialogues with children, there is an assignment of which utterances belong to a certain scenario. However, there are cases in which the same utterance exists in multiple scenario. In such cases, the dataset creators manually determined which scenario each utterance belonged to based on experimental records. For the child's response, each utterance was labeled according to the child's tension. The prosody of the child's speech was output as a feature using openSMILE [13]. We used these data to enable the proposed system to learn dialogue selections.

The second dataset is the Hazumi dataset [14], which served as the source domain data for deep learning to learn the polarity related to prosody and interest. Hazumi is a dataset of communication between adults and a system. We used the prosodic features and "IN_ternary" labels of the data in the Hazumi1712 dump file. Each utterance was labelled with the presence or absence of interest. The prosodic features were extracted using openSMILE [13]. In this study, we used these features to determine the existence of interest.

For supervised learning, we used the Hazumi dataset and the dataset of interactions with children. For unsupervised learning, only the dataset of interactions with children was used.

## 4 Experiment

The structure of the neural network is illustrated in Fig. 2, where the input is 384-dimensional prosodic information, and the output is a 3-dimensional soft-max layer. One child was included in the testing data at a time. These data were not used to adjust the model parameters. We used the testing data to provide rewards. Similar to a leave-one-out cross-validation, we ensured that all children were used in the testing data. The remaining children's data were used for training to avoid overfitting. One child was used for validation data, and the others were used for training data. The validation data and training data were separated using the holdout method. Overfitting was prevented by using the weights from the epochs in which the accuracy of the validation data was maximized. Pre-training on Hazumi's dataset was also performed to avoid overfitting. The number of epochs was adjusted using the leave-one-out cross-validation method. One person's data were used as the validation data and the other data were used as the training data.

For unsupervised learning, models trained on all children's data were used for clustering.

For prosodic information, we used the 384-dimensional features from openSMILE's INTERSPEECH 2009 Emotion Challenge feature set (IS09). The following rules were used to determine whether a child responded to a machine.

- If the child responded first after the start of the machine's speech, it was considered that the child responded.
- If the mother or the machine spoke first after the machine started speaking, it was considered that the child did not respond.

Additionally, if two or more types of utterances by the mother, machine, or child occurred within 0.14 s, they were considered to be stimulated [15].

- If both the child's and mother's utterances occur within 0.14 seconds of the system's utterance, then the child is assumed to have responded to the machine.
- If both the mother's and machine's utterances occur within 0.14 s of each other, then the child is considered to have responded to the later utterance.
- If both the child's and machine's utterances occur within 0.14 s, then we assume that the child may have responded to the previous machine utterance.

In this study, we calculated the expected value of rewards related to utterance selection in a scenario "Introduction".

### 4.1 Experiment1

The following experiments were conducted to compare the utterance selection learning by a mother, utterance selection learning with and without a child's response, and utterance selection learning using prosody.

1. The state was fixed, and the number of times each utterance was selected, speech choice learning with child responses, and speech choice learning with prosody were compared.
2. The number of times each utterance was then selected was compared. We used a fixed set of states, based on the presence or absence of a response and performed clustering by prosody.
3. We compared mother's speech choices, probabilities of speech choices in the presence or absence of responses as states and rewards, and the probabilities of speech choices when using prosody as states and rewards.

### 4.2 Experiment2

There are several methods for extracting states. For example, supervised learning can be used to change the number of classes. We used the information provided by prosody to define rewards and compared several methods based on states.

1. The k-means method is used to divide states (k = 2).
2. Separating states using the k-means method (k = 3).
3. Using teacher labels.
4. Using predictions by a neural network.

## 5 Result

In our experiments, we focused on a scenario called "Introduction". A comparison was made between the number of speech choices and probabilities of speech choices.

### 5.1 Experiment1

In all figures, the vertical axis presents the probability of utterance selection, and the horizontal axis represents the speech choices. In Fig. 3, the state is set to one. We compared the probabilities of the mother's speech choices, the probability of speech choice by reward with or without a response, and probabilities of speech choices by reward with or without a response and prosody. In Fig. 4, the states are divided as follows: one overall state, the state with the child's lack of response, the state with the child's response, and the state of the child's response (classified into two classes). Under these states, the probabilities of the mother's speech choices were compared. Figure 5 presents the following. We compare the mother's probabilities of the utterance selections (mother's selection), the probabilities of utterance selections in the presence or absence of a response as a state and reward (RL1), and the probabilities of utterance instructions when using the presence or absence of a response and prosody as a state and reward (RL2). For "mother's selection", the number of states is set to one.

When the number of states in the training data is one, all dialogue data is trained, but in scenarios where the machine is trained by separating states based on the presence or absence of a response or prosody, the first utterance is not trained because there is no previous machine utterance.

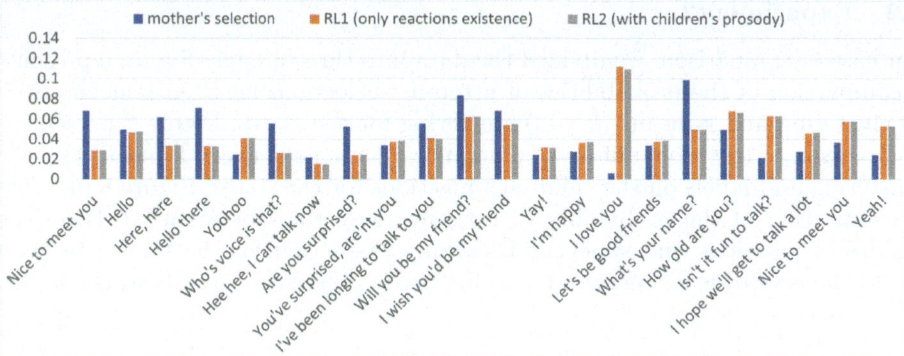


**Fig. 3.** Changes in the probabilities of utterance selections with rewards.

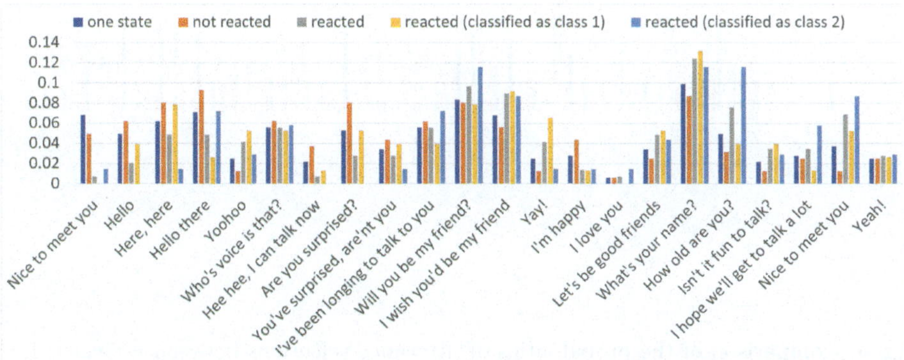


**Fig. 4.** Differences in the probability distribution of the mother's choice of statement depending on how the state is given.

**Fig. 5.** Comparison of mother's speech choices, probabilities of speech choices when using the presence or absence of a response as the state and reward, and probabilities of speech choices when using prosody as state and reward.

### 5.2 Experiment2

For ease of comparison, we divided the data into three groups. Figure 6 presents a comparison of the probabilities of utterance selections between k-means (k = 2) clustering and k-means (k = 3) clustering for the states. Figure 7 presents a comparison of the probabilities of utterance selections between k-means (k = 3) clustering and labels on the children's reactions for the states. Figure 8 presents a comparison of the probability of utterance selections between labels on the children's reactions and supervised learning prediction for the states. In both cases, the reward is a supervised learning prediction using prosody as the input.

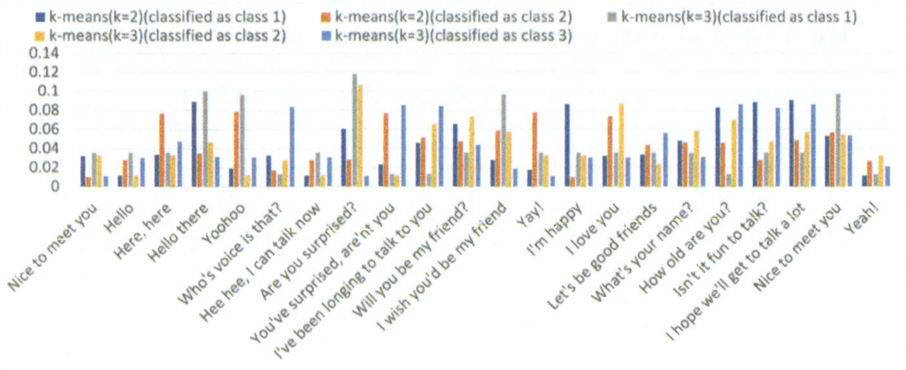


**Fig. 6.** Comparison of the probabilities of utterance selections between k-means (k = 2) clustering and k-means (k = 3) clustering for the states.

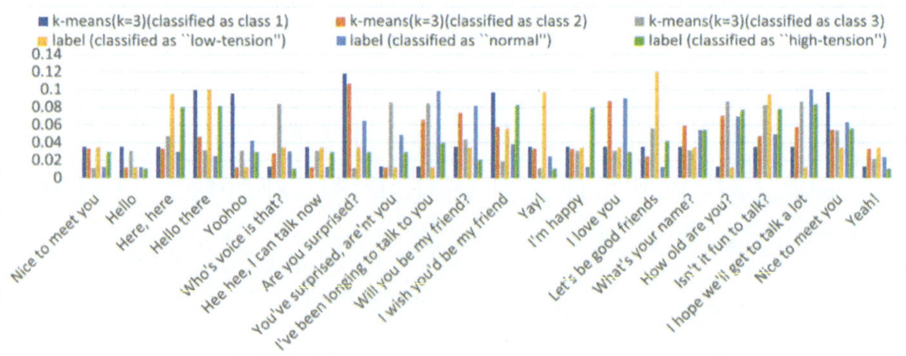


**Fig. 7.** Comparison of the probabilities of utterance selections between k-means (k = 3) clustering and labels on the children's reactions for the states.

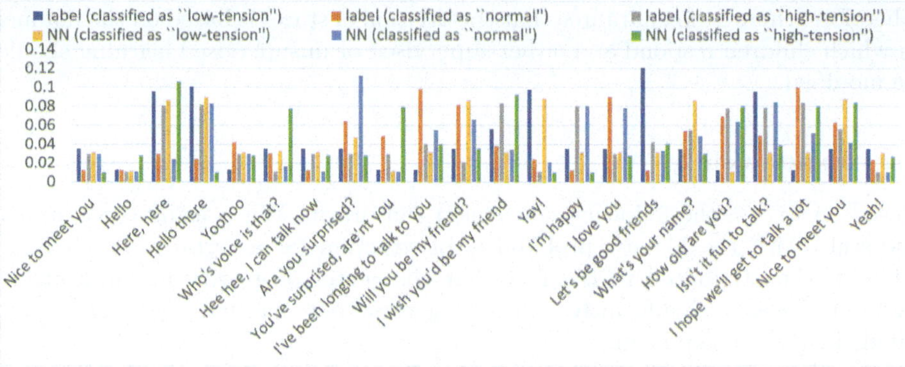


**Fig. 8.** Comparison of the probabilities of utterance selections between labels on the children's reactions and supervised learning (neural network) predictions for the states.

## 6 Discussion

### 6.1 Experiment1

Based on Fig. 3, the following conclusions can drawn. When prosody is used, as well as when the presence or absence of the child's response is used, the result differs from the mother's choice of utterance, but no major changes in utterance choices appear. Additionally, the distribution is flatter than when using the presence or absence of a response as a reward. This is because the neural network was trained by dividing the data into training data, validation data, and testing data to prevent overfitting, and the predictions were concentrated on "normal tension".

Based on Fig. 4, the following conclusion can drawn. The distribution of the mother's speech choice differs between the case with one state, case divided by the presence of a response, and case using prosody. Therefore, it is possible that the speech choice differs depending on the child's reaction to the previous speech choice. Therefore, it is possible that it is effective to divide states according to the child's reaction to the previous speech choice.

Based on Fig. 5, the following conclusion can drawn. Compared to Fig. 3, the probability distribution changes in some areas, even within the same state. This may be a result of the fact that the number of choices has been reduced by dividing the states and the number of responses for which the reward from the neural network is relatively large has increased. If it was possible to judge whether a child's response was favorable or unfavorable depending on the state, it was thought that when the children were divided into RL1 (reacted), RL2 (reacted (classified as 1)), and RL2 (reacted (classified as 2)), some speech choices would be higher and some would be lower. However, this was not the case.

Therefore, there is no guarantee that the division of states is based on the manner in which children respond and either supervised or unsupervised learning should be modified.

### 6.2 Experiment2

Based on Fig. 6, the following conclusion can drawn. When using three states, the state with the highest probability of selection for a certain line changes. This can be attributed to the fact that the center of gravity in the k-means method changed significantly, which suggests that the way the states were divided changed significantly.

Based on Fig. 7, the following conclusion can drawn. The probabilities of utterance selections were not observed to be consistent between the division of states by the k-means method and division of states using teacher labels. It can be concluded that the distribution of the labels and clustering of the k-means method do not match. Therefore, it can be said that the labels used in this study are characterized by the fact that accuracy is not guaranteed. It is necessary to investigate the validity of clustering and labels further.

Based on Fig. 8, the following conclusion can drawn. When the neural network predicts "normal-tension", it is possible to select more appropriate utterances by subdividing the states.

## 7 Future Works

First, there is a problem where the labels for children's responses used in our system may be inaccurate. It is necessary to improve the accuracy of teacher labels or to measure the accuracy of teacher labels in future research.

Additionally, the accuracy of the supervised and unsupervised learning methods used in this study is not guaranteed. Therefore, it is necessary to evaluate the accuracy of supervised and unsupervised learning using accurate labels.

In this study, the developed system did not learn utterance selections. Assumptions between different scenarios also need to be constructed. Therefore, it is necessary to construct a system that can learn dialogue, even when other scenarios are used.

Additionally, it is difficult to determine whether learned dialogues are appropriate because the current system uses graphs for comparison. However, in the data on dialogues with children, there is no guarantee that the mother's choice of utterance is the best choice. Therefore, it is necessary for a third party to judge whether a dialogue is appropriate for dialogue data.

## 8 Conclusion

In this paper, we proposed a method for learning utterance selections using prosodic information. In the proposed system, supervised learning is used for rewards and unsupervised learning is used for states. In our experiments, we compared utterance selections when using prosodic information to those when not using prosodic information. We also compared utterance selections for different methods of defining states. As a result, we determined that it is possible to use prosodic features and teacher labels for states to learn speech according to the target scenario.

**Acknowledgement.** This work was supported by JSPS KAKENHI Grant Numbers JP20H05564 and JP19K12081.

We would like to express our sincere thanks to Ms. Yukari Hori, Mr. Tetsuya Oyama, and Mr. Shinnosuke Furuya in Hakuhodo for their significant support in conducting this research.

## References

1. Hanada, S., Oka, N., Tanaka, K., Takahashi, T., Kubo, Y.: A study on automatic dialogue selection in a spoken dialogue system for children. In: Human-Agent Interaction Symposium (2021). (in Japanese)
2. Zhou, H., Huang, M., Zhang, T., Zhu, X., Liu, B.: Emotional chatting machine: emotional conversation generation with internal and external memory. arXiv e-prints, arXiv–1704 (2017)
3. Lin, Z., Madotto, A., Shin, J., Xu, P., Fung, P.: MoEL: mixture of empathetic listeners. In: Proceedings of the 2019 Conference on Empirical Methods in Natural Language Processing and the 9th International Joint Conference on Natural Language Processing (EMNLP-IJCNLP), pp. 121–132 (2019)
4. Majumder, N., Poria, S., Hazarika, D., Mihalcea, R., Gelbukh, A., Cambria, E.: Dialoguernn: An attentive rnn for emotion detection in conversations. In: Proceedings of the AAAI Conference on Artificial Intelligence, vol. 33, pp. 6818–6825 (2019)
5. Matsumoto, S., Araki, M.: Assesment of user's interests in multimodal dialog integrating multiple modalities. SIG-SLUD **5**(02), 86–87 (2018). (in Japanese)
6. Nishimoto, H., Komatani, K.: Predicting user's interest level in dialogues with multimodal features. In: Proceedings of the Annual Conference of JSAI, JSAI 2018, 3C2OS14b04 (2018). (in Japanese)
7. Horii, T., Nagai, Y., Asada, M.: Imitation of human expressions based on emotion estimation by mental simulation. Paladyn J. Behav. Robot. **7**(1), 40–54 (2016). De Gruyter
8. Nakayama, H.: Image feature extraction and transfer learning using deep convolutional neural networks. IEICE Tech. **115**(146), 55–59 (2015). (in Japanese)
9. Tan, C., Sun, F., Kong, T., Zhang, W., Yang, C., Liu, C.: A survey on deep transfer learning. In: Kůrková, V., Manolopoulos, Y., Hammer, B., Iliadis, L., Maglogiannis, I. (eds.) ICANN 2018. LNCS, vol. 11141, pp. 270–279. Springer, Cham (2018). https://doi.org/10.1007/978-3-030-01424-7_27
10. Huang, J.-T., Li, J., Yu, D., Deng, L., Gong, Y.: Cross-language knowledge transfer using multilingual deep neural network with shared hidden layers. In: 2013 IEEE International Conference on Acoustics, Speech and Signal Processing, pp. 7304–7308. IEEE (2013)

11. Mitsukuni, K., et al.: Analysis of the influence of mothers' traits and behaviors on children's conversational play with an utterance-output device. In: Proceedings of the 7th International Conference on Human-Agent Interaction, pp. 271–273 (2019)
12. Ichikawa, J., et al.: Investigation of interaction between children and an utterance-output device focused on personality. Trans. Jpn. Soc. Kansei Eng. **19**(2), 173–179 (2020). https://doi.org/10.5057/jjske.TJSKE-D-19-00058. (in Japanese)
13. Eyben, F., Wöllmer, M., Schuller, B.: Opensmile: the munich versatile and fast open-source audio feature extractor. In: Proceedings of the 18th ACM International Conference on Multimedia, pp. 1459–1462 (2010)
14. Araki, M., et al.: Collection of multimodal dialog data and analysis of the result of annotation of users' interests. SIG-SLUD **5**(02), 20–25 (2017). (in Japanese)
15. Oyama, T.: Historical background and the present status of reaction time studies. Jpn. J. Ergon. **21**(2), 57–64 (1985). (in Japanese)

# Facial Conditional Translation Using Multi-CartoonGAN with Central Biasing Instance Normalization

Rina Komatsu(✉) and Tad Gonsalves

Sophia University, Tokyo, Japan
r_komatsu@eagle.sophia.ac.jp, t-gonsal@sophia.ac.jp

**Abstract.** Image-to-Image Translation tasks sometimes need to deal with different appearance translations like real-world face to anime style. Existing studies try constructing robust models with many technical layers. However, the computational cost as well as the time required for training rapidly increases with the addition of layers. This study proposes N-to-N domains translation model named Multi-CartoonGAN which aims at learning difficult translation tasks and simulataneously keep the computational cost as low as possible. The Generator consists of a pre-trained VGG net and Central Biasing Normalization for stable training. The results show that the proposed model could perform not only facial emotion translation but also artistic face style translation including portrait, anime and emaki. Our model training also shows diminished computational cost compared to the existing models when dealing with difficult appearance translations.

**Keywords:** Image-to-Image translation · AI · Deep learning · GAN

## 1 Introduction

Image-to-Image translation using deep learning has a variety of tasks such as generating segmented maps, coloring from gray-scale images, and super-resolution. These translation models aim to learn the mapping to output images closer to the target domain using Generative Adversarial Networks (GAN) [1]. As the novel translation model with adversarial learning, CycleGAN [2] established one-to-one domain translation architecture without paired datasets by adopting adversarial loss and cycle consistency loss. U-GAT-IT [3] tried overcoming extreme appearance translation such as real-world face to anime face by using the normalization named Adaptive Instance-Layer Normalization (AdaLIN) and Class Activation Maps (CAM) logits for adversarial learning. StarGAN [4] utilizes the additional input of conditional vector and contributes to expanding the variation of translation by handling single Generator. StarGAN could establish N-to-N domains translation architecture.

Since there are studies concerning one-to-one domain translation for extreme appearance translation, we experimented with N-to-N domains translation model


Y. Takama et al. (Eds.): JSAI 2021, AISC 1423, pp. 57–68, 2022.
https://doi.org/10.1007/978-3-030-96451-1_6

with robust translation performance. However, just implementing technical methods to N-to-N translation is not an adequate solution. There is the other problem of computational cost when we start to train the model. If the model based on CycleGAN implements more technical layers and reuses the Generator to obtain cycle consistency loss, the computational cost of training on GPU will result in increased memory and time. To solve this trade-off between constructing robust N-to-N domains translation for extreme appearance translation and saving computational cost on GPU while training, this study proposes the model named **Multi-CartoonGAN**.

CartoonGAN [5] is a one-to-one domain translation model which employs pre-trained VGG net to calculate content loss by comparing extracted feature maps between the content and the translated images. We employed this model expecting the computational cost will be less because the step of reusing Generator is not needed. Our study developed CartoonGAN as multiple conditional translation model. Further, for handling conditional outputs and being robust to extreme appearance translation, our Multi-CartoonGAN employs Central Biasing Instance Normalization (CBIN) as conditional input in the Generator and adopts CAM logits in the Discriminator.

In our translation experiments, we tried training conditional translation using Multi-CartoonGAN for *facial attribute translation* experiment and *real-world face to artistic face translation* experiment. In the facial attribute translation, the translation models need to learn with 7 kinds of emotion categories such as happiness, sadness, anger, etc. In real-world face to artistic face translation, the models learned 3 kinds of facial art categories: portrait, anime and emaki. This translation requires extreme appearance translation especially translating into anime and emaki styles.

## 2 Related Studies

### 2.1 One-to-One Domain Translation

The aim of One-to-One domains translation task is translating the content image $x$ which belongs to domain $X$ to output image $G(x)$ which seems likely belonging to the target domain $Y$ through adversarial learning by Generator $G$ and Discriminator $D$.

Adversarial learning has been adapted to translation architectures. pix2pix [6] employs adversarial loss to make the output close to the real. However, this architecture requires paired datasets to obtain the Ll distance. CycleGAN [2] could solve this problem by adopting cycle consistency loss using two Generators: $G_X$ and $G_Y$. Cycle consistency loss is calculated by comparing L1 distance between the content image $x$ and the reconstructed image $G_X(G_Y(x))$. Adversarial loss is calculated by exclusive Discriminators $D_X$ and $D_Y$ for each domain.

Kim et al. [3] point out that the image-to-image translation methods succeed in handling local appearance changes like photo to art, but they typically fail in handling extreme appearance changes like dog to cat. U-GAT-It tried solving the difficulty of extreme appearance translation by adapting AdaLIN to help the model to flexibly control the amount of change that the target domain requires. TransGaGa [7] builts the mapping about appearance-geometry space between two domains and exchanges geometry features and appearance ones. And NICE-GAN [8] reuses Discriminators as encoder to directly train through adversarial loss and makes the plug-in encoder more informative.

One-to-one domain translation models have the limitation when we want to output $N$ styles of translated images. To obtain $N$ kinds of translated images, we need to train $N$ pairs of Generator and Discriminator. It takes a lot of time to complete translating training with all models. Further, robust translation models like U-GAT-IT result in large computational cost because of employing technical layers and reusing the Generator to calculate cycle consistency loss.

### 2.2 N-to-N Domains Translation

To be able to control the style of output, conditional input which relates with targeted style is utilized. Mizra and Osindero [9] proposed the conditional Generative Adversarial Nets (cGAN) architecture which trains the Generator to be able to output conditional images by simply feeding the conditional inputs with vector representation. This method is adapted in image-to-image translation models to control outputting the style of translation.

IcGAN [11] and StarGAN [4] employ conditional translation training by in putting content image and conditional input with vector representation to the Generator to be able to obtain conditional translated image following conditional input. The conditional input is directly injected into the Generator. However, Yu et al. [10] pointed out that injecting method leads to mode collapse because of normalization strategies like instance normalization which might cause inconsistency of feature distribution. Instead of injecting conditional input, Yu et al. [10]proposed Central Biasing Normalization (CBN) which deals with conditional input as bias, and the bias is added to normalized input. GatedGAN [12] implements $N$ branches of residual blocks which are placed between the encoder and the decoder, and outputs conditional translated image by selecting the branch. ComboGAN [13] combines encoder and selected decoder which relates to the target domain. The existing studies showed good performance in local appearance translation such as facial attribution, and photo to art. Our study deals with extreme appearance translations.

## 3 Proposed Model: Multi-CartoonGAN

### 3.1 Overview of Our Model

Figure 1 shows the overview of Generator and Discriminator in our Multi-CartoonGAN. In the Generator, we implemented CBIN in the encoder, following SingleGAN method [14] which employs CBIN developed from CBN. In bottleneck between encoder and decoder, there are 2 kinds of residual blocks: residual blocks with CBIN and the ones with AdaLIN [3]. When the feature maps input

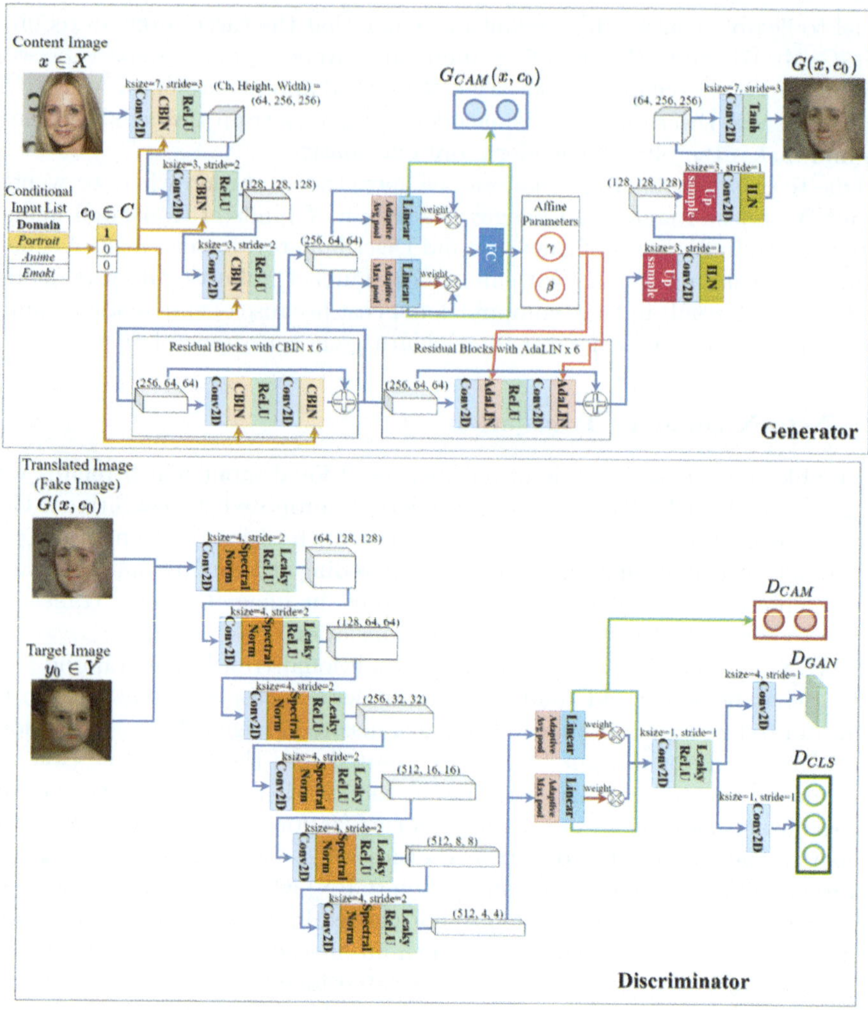


**Fig. 1.** Overview of our Multi-CartoonGAN

into residual blocks with AdaLIN, affine parameters are calculated through adaptive pool layers. To obtain which area needs to be changed using CAM logits, the adversarial criterion $G_{CAM}$ is output.

In the Discriminator, we employed spectral normalization [15] after convolution to help the Discriminator to maintain stability during training. There are 3 kinds of outputs: $D_{CAM}$ for discriminating attention feature maps from CAM logits, $D_{GAN}$ for discriminating from patches of feature maps, and $D_{CLS}$ to train the Discriminator to classify what domain the input image belongs to.

### 3.2 Full Objective Loss

Equations (1)–(9) are the loss functions we used for conditional translation training.

Equation (1) is the full objective loss for our Generator $G$. This equation consists of 5 contents: Eq. (3) is the adversarial loss using patches from $D$ based on LSGAN [16] which employs L2 norm to determine whether the input data belongs to $real$ or $fake$, Eq. (4) is the adversarial loss using CAM logits by LSGAN, Eq. (5) is the adversarial loss by discriminating CAM logits from $G_{CAM}$ using Binary Cross Entropy, Eq. (8) is content loss using conv4_4 layer from pre-trained VGG19 net following CartoonGAN method [5], and Eq. (6) is the classifying loss whether the translated image with $c$ could be judged as belonging to domain $c$ through $D_{CLS}$ like AC-GAN [17].

Equation (2) is the full objective loss for our Discriminator $D$. This equation consists of 4 contents: Eq. (3), Eq. (4), Eq. (7) is the classification loss that $D$ could classify the target domain, Eq. (9) is the gradient penalty from WGAN-GP [18] for making the training of GAN stable.

As for the hyperparameters for training, we set $\lambda_{GAN} = 3.0$, $\lambda_{CAM}^{G} = 1000.0$, $\lambda_{VGG} = 1.0$ and $\lambda_{GP} = 10.0$.

$$L_G = \lambda_{GAN} L_{GAN}(G, D) + L_{CAM}^{D}(G, D) + \lambda_{CAM}^{G} L_{CAM}^{G} + \lambda_{VGG} L_{VGG} + L_{CLS}^{G} \tag{1}$$

$$L_D = \lambda_{GAN} L_{GAN}(G, D) + L_{CAM}^{D}(G, D) + L_{CLS}^{D} + \lambda_{GP} L_{GP} \tag{2}$$

$$\begin{aligned} L_{GAN}(G, D) = {} & \mathbb{E}_{y \sim p_{data}(y)}[(D(y))^2] + \\ & \mathbb{E}_{y_{blur} \sim p_{data}(y_{blur})}[(1 - D(y_{blur}))^2] + \mathbb{E}_{x \sim p_{data}(x), c}[(1 - D(G(x, c)))^2] \end{aligned} \tag{3}$$

$$\begin{aligned} L_{CAM}^{D}(G, D) = {} & \mathbb{E}_{y \sim p_{data}(y)}[(D_{CAM}(y))^2] + \\ & \mathbb{E}_{y_{blur} \sim p_{data}(y_{blur})}[(1 - D_{CAM}(y_{blur}))^2] + \mathbb{E}_{x \sim p_{data}(x), c}[(1 - D_{CAM}(G(x, c)))^2] \end{aligned} \tag{4}$$

$$L_{CAM}^{G}(G) = -(\mathbb{E}_{x \sim p_{data}(x), c}[log(1 - G_{CAM}(x, c))]) \tag{5}$$

$$L_{CLS}^{G} = \mathbb{E}_{x \sim p_{data}(x), c}[-log(D_{CLS}(c|G(x, c)))] \tag{6}$$

$$L_{CLS}^{D} = \mathbb{E}_{y \sim p_{data}(y), c}[-log(D_{CLS}(c|y)] \tag{7}$$

$$L_{VGG} = \mathbb{E}_{x \sim p_{data}(x), c}||F_{conv4\_4}(G(x, c)) - F_{conv4\_4}(x)||_1 \tag{8}$$

$$L_{GP} = \mathbb{E}_{\hat{x}}[(||\nabla_{\hat{x}} D(\hat{x})||_2 - 1)^2] \tag{9}$$

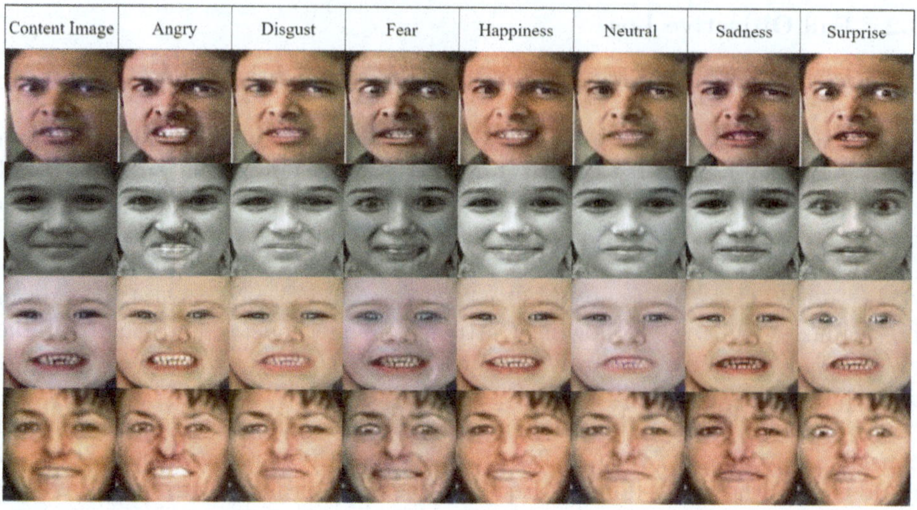


**Fig. 2.** Results of conditional facial emotion translation using trained Multi-CartoonGAN

# 4 Experimental Setup

## 4.1 Experiment 1: Facial Emotion Translation

In this experiment, we trained N-to-N domains translation models: StarGAN, SingleGAN, and Multi-CartoonGAN on the dataset RAF-DB Basic [20]. From this dataset, we used each of the face images with 7 kinds of attributes: *Anger, Disgust, Fear, Happiness, Neutral, Sadness and Surprise.*

## 4.2 Experiment 2: Artistic Face Translation

In this experiment, we trained CycleGAN, U-GAT-IT, NICE-GAN, CartoonGAN, StarGAN, SingleGAN and Multi-CartoonGAN. As the content domain, we employed real-world face image dataset CelebA-HQ [21]. As the 3 kinds of artistic face domains, we considered the domains *Portrait, Anime and Emaki.* We employed portrait images from MetFaces Dataset [22], anime face images from U-GAT-IT's demonstration for selfie to anime translation [3] and emaki images from Kaokore Dataset [23].

## 4.3 Training Setting

We constructed Multi-CartoonGAN and implemented conditional translation training using deep learning library Pytorch. In training, we employed function Adam [24] for optimizing $G$ and $D$ setting hyperparameters $\beta_1 = 0.5$ and $\beta_2 = 0.999$ with learning rate $lr = 0.0001$ and $weightdecay = 0.0001$. For completing the training as fast as possible, we set batch size =15 and utilized GPU: NVIDIA TITAN RTX GPU.

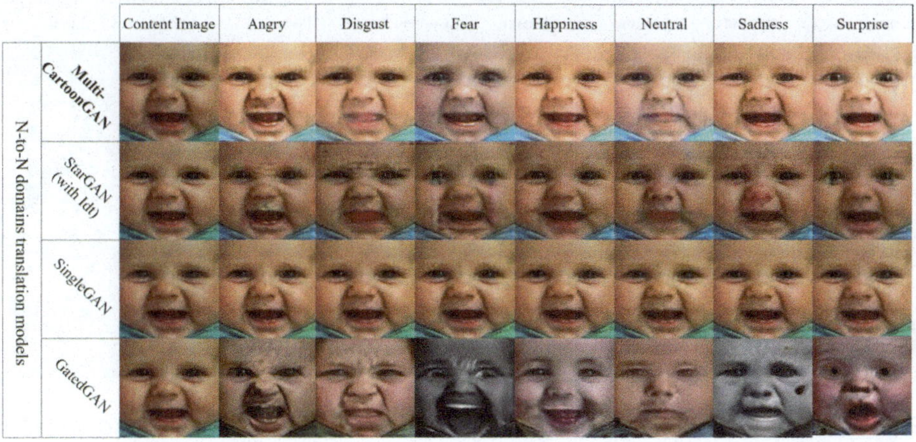


**Fig. 3.** Comparing results of conditional facial emotion translation among N-to-N domains translation models

# 5 Results

## 5.1 Visualized Results

**Experiment 1: Facial Emotion Translation.** We input the face images from testing dataset of RAF-DB to trained domain translation Generator. Figure 2 shows the results which are translated by trained Multi-CartoonGAN. We can infer from the translated images that our model could change facial features such as eyes, eyebrows, and winkles from the content images following targeted emotion labels without asking for gender and age. For instance, each of the face images which are translated as "angry" shows raised eyebrows and deep winkles for anger. The images which are translated as "happy" show gentle look showing teeth with smile and raised mouth angle. "Surprise" results, have characteristic open eyes. However, in some cases, our model translated needless area seeing the coloring features of the content image. Figure 3 compares conditional translated results using trained N-to-N domains translation models: our Multi-CartoonGAN, StarGAN with idt (which employs identity mapping loss to prevent translated image from extreme color changing), SingleGAN and GatedGAN. As each translated face shows, our proposed model, StarGAN and GatedGAN could output translated images with different facial features following the conditional input. SingleGAN could not show the differences among translated images and could not reflect the facial features the target domain has. GatedGAN was able to output images with extreme characteristic facial features, however, the physiognomy in some results was also extremely changed. In conclusion, our model could translate at a moderate level evaluating from the fact that the facial features of the content image were preserved, and that it could translate only the area necessary for expressing facial emotions.

**Fig. 4.** Results of conditional artistic face translation using trained Multi-CartoonGAN

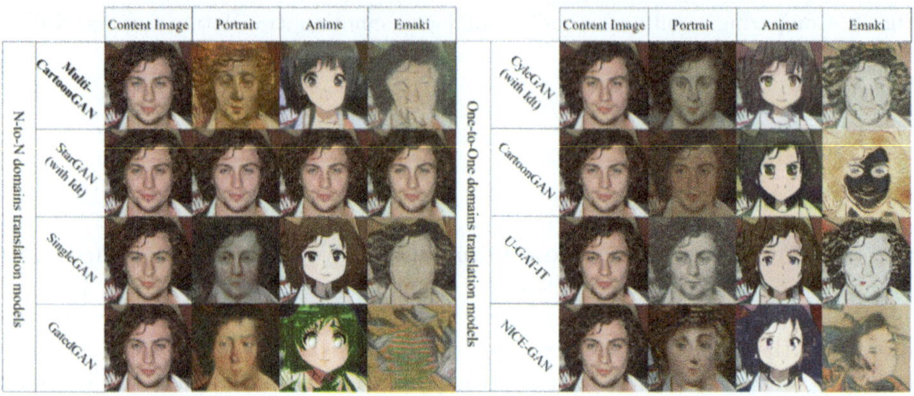


**Fig. 5.** Comparing results of conditional artistic face translation among image-to-image translation models

**Experiment 2: Artistic Face Translation.** We translated the test face images to 3 kinds of artistic styles via the trained Generator. Figure 4 shows the conditional translated results using trained Multi-CartoonGAN. Viewing each of the translated images from the content images, these images could have different and distinctive artistic features. For instance, translated "Portrait" images had the unique brush stroke for describing face shape and shade. "Anime" images, as the anime dataset mainly collected anime girl faces, the output images (translated without considering gender) tended to have long hair and small faces. Also, these images tried imitating anime style coloring such as clear edge and glossy hair. Finally, the translated "Emaki" images could describe characteristic edges and unique light color like the Asian ink brush.

Figure 5 shows comparative conditional translated results using trained One-to One domain translation model: CycleGAN with Idt (which employs identity mapping loss) CartoonGAN, U-GAT-IT and NICE-GAN, and N-to-N domains translation models: Multi-CartoonGAN, StarGAN with Idt, SingleGAN and GatedGAN. Viewing each of the output style images, StarGAN with Idt could not output any difference from the content image. Viewing the output images with portrait style in each model, these models could output likely portrait style evaluating from the unique brush stroke. Almost all the translated anime style images could have characteristic coloring with the exception of CartoonGAN which outputs unnaturally big eye designs so much so that discriminating the edge line of each eye is not easy. As for the translated emaki style images, GatedGAN, CartoonGAN and NICE-GAN tended to lose facial features from the content image. Our model could select likely Asian ink brush coloring; however, the strength of emaki edge features was weaker than CycleGAN with Idt and U-GAT-IT.

**Table 1.** Comparing Computational Cost on GPU (measured in Artistic Face Translation)

| Model name | Time (sec) | MiB |
|---|---|---|
| **Multi-CartoonGAN** | 8122.056 | 3388 |
| CycleGAN with Idt | 12455.54(per domain)x3 | 3947 |
| CartoonGAN | 7105.51(per domain)x3 | 1902 |
| U-GAT-IT | 17182.95(per domain)x3 | 7693 |
| NICE-GAN | 18866.45(per domain)x3 | 19952 |
| StarGAN with Idt | 11468.74 | 3198 |
| SingleGAN | 10053.22 | 4159 |
| GatedGAN | 5725.86 | 1857 |

## 5.2 Comparing Computational Cost on GPU

To investigate the computational cost on GPU and the time to complete 1 epoch training, we measured them in Artistic Face Translation experiment setting batch size=l. Table 1 shows the comparative results among the various image-to-image translation models.

Since One-to-One domain translation models needed to train for each domain, they took a lot of time to perform all translations. Also, the robust translation models U-GAT-IT and NICE-GAN had higher computational cost because of implementing technical layers and reconstruction process for cycle consistency loss. On the other hand, our model showed reduced computational cost than the existing robust translation models even after implementing technical layers of residual blocks with AdaLIN. We suggested utilizing VGG19 for content loss since it solved the computational cost problem in CartoonGAN implementation. Our method also took less time to complete 1 epoch training than the existing N-to-N domains translation models, StarGAN with Idt and SingleGAN.

**Table 2.** Comparing KIDx100 ± stdx100 Results in Artistic Face Translation

| Model name | Portrait | Anime | Emaki |
|---|---|---|---|
| **Multi-CartoonGAN** | 2.33 ± 0.06 | 7.69 ± 0.18 | 14.86 ± 0.24 |
| CycleGAN with Idt | **1.33 ± 0.04** | 2.85 ± 0.07 | 8.80 ± 0.19 |
| CartoonGAN | 4.77 ± 0.09 | 6.89 ± 0.09 | 17.86 ± 0.28 |
| U-GAT-IT | 1.40 ± 0.06 | **1.82 ± 0.05** | 8.40 ± 0.22 |
| NICE-GAN | 3.31 ± 0.10 | 2.66 ± 0.06 | **4.21 ± 0.12** |
| StarGAN with Idt | 10.32 ± 0.16 | 28.28 ± 0.22 | 27.43 ± 0.32 |
| SingleGAN | 1.72 ± 0.07 | 3.13 ± 0.08 | 8.42 ± 0.19 |
| GatedGAN | 6.78 ± 0.09 | 13.21 ± 0.14 | 12.81 ± 0.14 |

## 6 Conclusion

We introduced a new N-to-N domains translation model named "Multi-CartoonGAN" to handle conditional N domains translation by implementing CBIN and adapting it to the extreme appearance translation by employing AdaLIN. Using this model, we tried solving the trade-off problem of making the model robust by implementing more technical layers and reusing the Generator for reconstruction. As shown in our visualizing result section, our trained Multi-CartoonGAN could perform conditional translation following conditional input and could reflect the unique features the targeted domain has. Also, as showing in the result section of comparing the computational cost on GPU, our proposed model could save lower the computational cost compared to the existing robust translation models such as U-GAT-IT and NICEGAN.

Also, in the visualized result section of Artistic Face Translation, we found some limitations. The design of translation variety tended to be less considering from the design of eyes in each of the translated anime images. Secondly, the strength of the features in each artistic face image was not stronger than CycleGAN with ldt and U-GAT-IT, especially in translating into emaki style. We estimated the quality of each translation with value by using KID [25] which computes the maximum mean discrepancy between two Inception representations of target domain images and translated images. Lower score means the translated images could have the same characteristic features the target images have. Table 2 shows the comparative result in Artistic Face Translation experiment. Lower score is represented with bold text. Viewing the table result, our model was not able to achieve lower score than the other One-to-One domain translation models. Apparantly, the cause of the lower scores is the lack of translation variety.

To further improve our model, we would like to implement the following items:

- To increase the translation variety, we will further develop our model into Multi-Modal Translation model by injecting random noise (latent code) or targeted style image as conditional input.
- To enhance the strength of features for each target domain, we may need to modify the construction of the Discriminator by adapting conditional normalization function whose affine parameters are controlled by the conditional input.

## References


1. Goodfellow, I.J., et al.: Generative adversarial nets. In: Proceedings of the 27th International Conference on Neural Information Processing Systems, NIPS 2014, vol. 2, pp. 2672–2680 (2014)
2. Zhu, J.-Y., Park, T., Isola, P., Efros, A.A.: Unpaired image-to-image translation using cycle-consistent adversarial networks. In: Proceedings of the IEEE International Conference on Computer Vision, pp. 2223–2232 (2017)
3. Kim, J., Kim, M., Kang, H., Lee, K.: U-GAT-IT: unsupervised generatove attentional networks with adaptive layer instance normalization for image-to-image translation. In: International Conference on Learning Representations 2020, pp. 1–19 (2019)
4. Choi, Y., Choi, M., Kim, M., Ha, J.-W., Kim, S., Choo, J.: StarGAN: unified generative adversarial networks for multi-domain image-to-image translation. In: Proceedings of the IEEE Conference on Computer Vision and Pattern Recognition, pp. 8789–8797 (2018)
5. Chen, Y., Lai, Y.-K., Liu, Y.-J.: CartoonGAN: generative adversarial networks for photo cartoonization. In: Proceedings of the IEEE Conference on Computer Vision and Pattern Recognition, pp. 9465–9474 (2018)
6. Isola, P., Zhu, J.-Y., Zhou, T., Efros, A.A.: Image-to-image translation with conditional adversarial networks. In: Proceedings of the IEEE Conference on Computer Vision and Pattern Recognition, pp. 1125–1134 (2017)
7. Wu, W., Cao, K., Li, C., Qian, C., Loy, C.C.: TransGaGa: geometry-aware unsupervised image-to-image translation. In: Proceedings of the IEEE/CVF Conference on Computer Vision and Pattern Recognition, pp. 8012–8021 (2019)
8. Chen, R., Huang, W., Huang, B., Sun, F., Fang, B.: Reusing discriminators for encoding: towards unsupervised image-to-image translation. In: Proceedings of the IEEE/CVF Conference on Computer Vision and Pattern Recognition, pp. 8168–8177 (2020)
9. Mizra, M., Osindero, S.: Conditional generative adversarial nets. arXiv preprint arXiv:1411.1784 (2014)
10. Yu, X., Ying, Z., Li, T., Liu, S., Li, G.: Multi-mapping image-to-image translation with central biasing normalization. arXiv preprint arXiv:1806.10050 (2018)
11. Perarnau, G., van de Weijer, J., Raducanu, B., Álvarez, J.M.: Invertible conditional GANs for image editing. In: NIPS 2016 Workshop on Adversarial Training, pp. 1–9 (2016)

12. Chen, X., Xu, C., Yang, X., Song, L., Tao, D.: Gated-GAN: adversarial gated networks for multi-collection style transfer. IEEE Trans. Image Process. **28**(2), 546–560 (2018). https://doi.org/10.1109/TIP.2018.2869695
13. Anoosheh, A., Agustsson, E., Timofte, R., Gool, L.V.: ComboGAN: unrestrained scalability for image domain translation. In: Proceedings of the IEEE Conference on Computer Vision and Pattern Recognition Workshops, pp. 783–790 (2018)
14. Yu, X., Cai, X., Ying, Z., Li, T., Li, G.: SingleGAN: image-to-image translation by a single-generator network using multiple generative adversarial learning. In: Jawahar, C.V., Li, H., Mori, G., Schindler, K. (eds.) ACCV 2018. LNCS, vol. 11365, pp. 341–356. Springer, Cham (2019). https://doi.org/10.1007/978-3-030-20873-8_22
15. Miyato, S., Kataoka, T., Koyama, M., Yoshida, Y.: Spectral normalization for generative adversarial networks. In: International Conference on Learning Representations 2018, pp. 1–26 (2018)
16. Mao, X., Li, Q., Xie, H., Lau, R.Y., Wang, Z., Paul Smolley, S.: Least squares generative adversarial networks. In: Proceedings of the IEEE International Conference on Computer Vision, pp. 2794–2802 (2017)
17. Odena, A., Olah, C., Shlens, J.: Conditional image synthesis with auxiliary classifier GANs. In: International Conference on Machine Learning, PMLR, pp. 2642–2651 (2017)
18. Gulrajani, I., Ahmed, F., Arjovsky, M., Dumoulin, V., Courville, A.: Improved training of Wasserstein GANs. In: Proceedings of the 31st International Conference on Neural Information Processing Systems, NIPS 2017, pp. 5769–5779 (2017)
19. Lee, C.-H., Liu, Z., Wu, L., Luo, P.: MaskGAN: towards diverse and interactive facial image manipulation. In: Proceedings of the IEEE/CVF Conference on Computer Vision and Pattern Recognition, pp. 5549–5558 (2020)
20. Li, S., Deng, W., Du, J.: Reliable crowdsourcing and deep locality-preserving learning for expression recognition in the wild. In: Proceedings of the IEEE Conference on Computer Vision and Pattern Recognition, pp. 2852–2861 (2017)
21. Karras, T., Aila, T., Laine, S., Lehtinen, J.: Progressive growing of GANs for improved quality, stability and variation. In: International Conference on Learning Representations 2018, pp. 1–26 (2018)
22. Karras, T., Aittala, M., Hellsten, J., Laine, S., Lehtinen, J., Aila, T.: Training generative adversarial networks with limited data. In: Advances in Neural Information Processing Systems 33 (NeurIPS 2020), pp. 1–11 (2020)
23. Tian, Y., Suzuki, C., Clanuwat, T., Bober-Irizar, M., Lamb, A., Kitamoto, A.: KaoKore: a pre-modern Japanese art facial expression dataset. arXiv preprint arXiv:2002.08595 (2020). 10.20676/0000035
24. Kingma, D.P., Ba, J.L.: Adam: a method for stochastic optimization. In: International Conference for Learning Representations 2015, pp. 1–13 (2015)
25. Sutherland, J.D., Arbel, M., Gretton, A.: Demystifying MMD GANs. In: International Conference for Learning Representations 2018, pp. 1–36 (2018)

# Styled Comic Portrait Synthesis Based on GAN

Yen-Chia Chen[1](✉), Lieu-Hen Chen[2], Hiroki Shibata[1], and Yasufumi Takama[1]

[1] Tokyo Metropolitan University, Hachioji, Japan
[2] National Chi Nan University, Nantou County, Taiwan

**Abstract.** Although there are many studies on NPR (Non-Photorealistic Rendering) image synthesis using GANs (Generative Adversarial Networks), it is still difficult to create high-quality comic portraits of a real person. Moreover, there are few studies focused on the painting styles of comic authors, which is the style that makes a comic visually unique. It is expected that for comic readers, the synthesized comic portraits can be more attractive and meaningful if the portraits are presented in the user-preferred comic styles. Therefore, this paper proposes a styled comic portraits synthesis system based on CycleGAN and PIX2PIX. By integrating Deep Learning and NPR techniques, the proposed system aims to transform user's real pictures into comic portraits with features preserved and defined painting style presented. CNN (Convolutional Neural Networks) is trained to classify the painting styles of comic authors. After that, two sets of GANs are trained with classified and augmented dataset, which is generated by mapping comic characters' 2D texture onto perturbed and deformed 3D facial models. The experiment results show that the proposed system can successfully create clear and vivid comic portraits, which has a great potential to serve as a useful tool for social network and comic industry.

**Keywords:** GAN · Comics · Painting styles · Portraits

## 1 Introduction

There is a long history for human beings to use graphic icons, such as emblems and drawings, to represent their identity. After the photographic technology was developed in the 19th century, photos became commonly used for the above mentioned purpose. Today, in the information era, tremendous amounts of image data including portrait photos are created every day because of the popularization of smartphones and other mobile devices. These portrait photos are very frequently used for different purposes on various occasions and places including social networks. With the increasing penetration of social networks across the world, the two contradictory issues of openness and privacy for personal information have both become more important. To solve this contradiction, one possible approach is to adopt comic-style portraits which preserves the user's facial features, instead of using real face images.

However, it is still difficult to automatically create high-quality comic portraits of a real person, even though there are many researches on NPR (Non-Photorealistic Rendering) image synthesis using GANs (Generative Adversarial Networks). Moreover, there


Y. Takama et al. (Eds.): JSAI 2021, AISC 1423, pp. 69–80, 2022.
https://doi.org/10.1007/978-3-030-96451-1_7

are few studies focused on representing the painting styles of comic authors, although most of the famous comic authors have their own painting styles, which make their works very distinguishable and valuable. It is expected that for comic readers, the synthesized comic portraits can be more attractive and meaningful if these portraits are presented in the user-preferred comic styles.

This paper proposes a styled comic portraits synthesis system based on CycleGAN [1–3] and PIX2PIX [4]. By integrating Deep Learning and NPR techniques, the proposed system aims to transform user's real pictures into comic portraits with real features preserved and user defined painting style presented simultaneously. To achieve this goal, a CNN (Convolutional Neural Networks) is trained to classify 6 famous Japanese comic authors. The classification result is used to quantify the uniqueness of comic painting style through the confusion matrix. After that, CycleGAN-based system is trained using the classified and augmented dataset, which is generated by mapping comic characters' 2D texture onto perturbed and deformed 3D facial models. In addition, PIX2PIX-based system is trained to transform the result by the CycleGAN into styled comic portrait. The effectiveness of the proposed system is shown with the examples of synthesizing comic portraits that incorporate three different comic authors.

The remaining part of this paper is organized as follows: Sect. 2 briefly discusses related work. Section 3 describes the proposed system and its implementation. Experimental results are discussed in Sect. 4. Note that in this paper, comics are used for research use only.

## 2 Related Work

### 2.1 Generative Adversarial Networks

In the research of Deep Learning such as GANs, image synthesis has always been an important topic. There are several published studies, including StackGAN [5, 6], DCGAN [7], and CycleGAN, which have demonstrated good results for different purposes. This subsection briefly explains typical types of GANs:

**StackGAN.** StackGAN uses text to generate images that match the description of given articles. The system architecture of StackGAN can be divided into two parts. The first part aims to understand the meaning of the text and produce fuzzy images that match the narrative descriptions. The second part aims to make the output images more real and natural by further refining the blurred images. The previous methods generate pictures from text mainly by retrieving the related pictures from the database according to the extracted keywords. In contrast, StackGAN tries to directly understand the meaning of the article and then generate related images. Compared to traditional GANs, StackGAN can generate more natural pictures because it adopts more parameters (of tanh and Leaky ReLU) and batch normalization to deal with the situation of unstable learning.

**DCGAN.** DCGAN generates pictures that are similar to the training datasets. The main difference of StackGAN and DCGAN is that the latter uses convolution layers and deconvolution layers instead of the fully connected layers to mitigate the problem that the fully connected and pooling layers tend to cause the loss of picture details. Similar to Word2Vec in natural language processing, DCGAN can perform logical addition and subtraction operations in terms of various elements of an image. For example, it is shown in [7] that a smiling face is synthesized by adding a smiling element, which is extracted from other pictures of smiling face, to a face image with no expression.

**CycleGAN.** CycleGAN converts the surface textures of the target object in a picture into different one in other pictures. It is shown in [2] that the skin of a horse in a picture are converted to that of zebra. The biggest advantage of CycleGAN is that it can be trained without the pair of images as the example of the conversion. For training deep learning models, in general, it is very difficult to prepare training and test dataset. However, CycleGAN can randomly pick up two input images from input datasets, and convert the surface texture of one input image to another one.

### 2.2 NPR and Comic Computing

Different from the conventional research topics in the field of Computer Graphics, Non-Photorealistic Rendering (NPR) focuses on rendering images by emulating the visual effects of hand drawing. For example, Chen etc. have proposed a 3D mesh approach which generates the thickness of strokes to represent the volumetric effects on canvas [8]. Moreover, NPR techniques are also used to render images in different artistic drawing styles such as the Impressionism of oil paintings [9]. In that paper, real images such as street views are retouched with different weather effects by applying NPR techniques to sky regions that are segmented by using a deep learning method.

As shown in these previous works, NPR techniques can be used to synthesize cartoon-like pictures from real world images in a very convenient and effective manner. Therefore, NPR techniques are well adopted by Comic Computing, which is a new research topic involving various fields such as artificial intelligence and computer graphics, as important tools for rendering comic pictures [10, 11].

## 3 Proposed System and Its Implementation

The flow of the proposed system is as follows (Fig. 1):

1. At first, a CNN is trained for automatic classification of comic styles. A classifier is trained so that comic characters' facial images of the same comic author can be classified into the same class. After that, one author (class) $a$ is selected and those images are regarded as Target domain $Y(a)$.
2. Grayscale images in $Y(a)$ are texture-mapped onto 3D facial models, which are perturbed and deformed in advance to represent different types of face shapes. These 3D models with comic texture are then used to generate reality facial images for artificially augmentation. All these generated images are stored in another group named Source Domain $X(a)$.
3. Images in Target Domain $Y(a)$ and Source Domain $X(a)$ are then used to train the CycleGAN, in which PatchGAN [12, 13], the original discriminator of the CycleGAN, is replaced with a five-layer CNN.

4. Grayscale comic images are converted into colored comic images.
5. Using the CycleGAN, colored comic images generated in Step 4 are input to a PIX2PIX-based GAN, in which a discriminator is replaced with a five-layer CNN. This PIX2PIX-based GAN is trained to synthesize grayscale comic images.
6. Real portrait photos are input to the PIX2PIX-based GAN to synthesize comic portraits as an output.

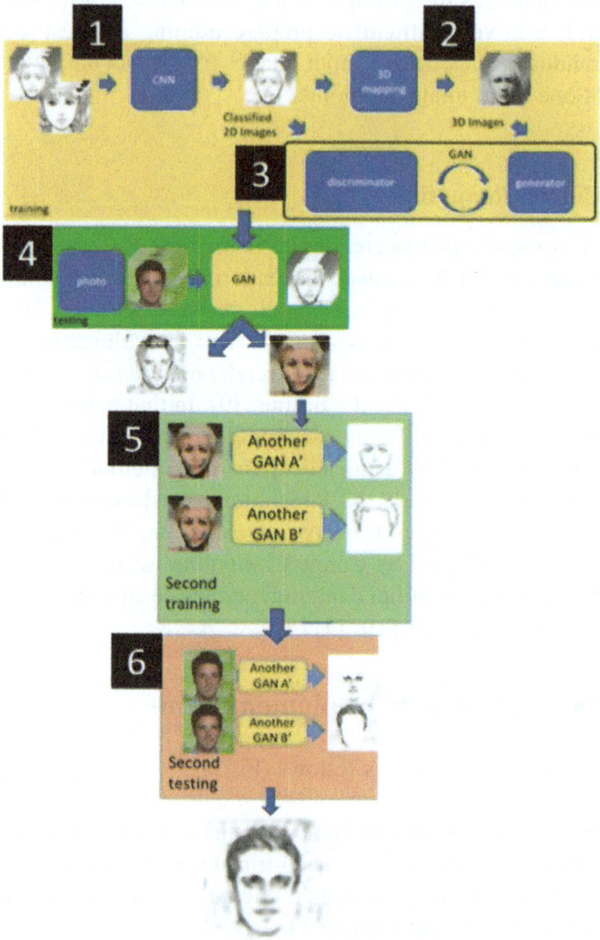


**Fig. 1.** Flowchart of proposed system.

### 3.1 CNN for Comic Style Classification

1,945 comic facial images from six comic works by different Japanese comic authors were collected as training dataset, including 363 Eiichiro Oda's "One Piece", 330 Taito

Kubo's "BLEACH", 320 Gosho Aoyama's "Conan", 343 Masashi Kishimoto's "Naruto", 349 Fumiya Sato's "Kindaichi", and 240 Hideaki Sorachi's "Gintama". These images are used to train a five-layer CNN for style classification in Step 1.

### 3.2 Data Augmentation

As explained in Step 2, instead of using meaningless random noise, data augmentation is performed by texture-mapping original images onto 3D models to create distorted new images [14–16]. A 3D facial model is created with 1,096 vertex and 864 surfaces by using Blender, which is a free and open source for 3D model creation. In addition, this default model is deformed to represent different faces shapes. Then, 2D comic images are mapped onto these 3D face model. Finally, as shown in Fig. 2, new images are generated by rendering the face models with comic textures for the purpose of data augmentation of reality facial images.

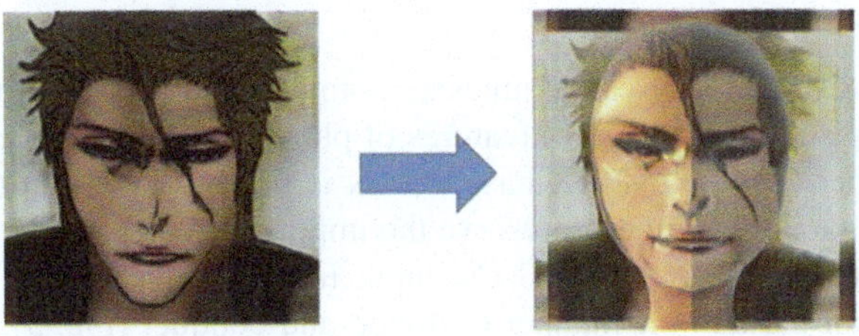

**Fig. 2.** Data augmentation by texture mapping (colored).

### 3.3 GAN for NPR Image Synthesis

As mentioned above, a CycleGAN is trained to convert grayscale comic pictures into colored images. It consists of two GANs, $G_a$ and $G_b$, which uses comic images and the photos as input respectively. It is practically beneficial because it is impossible to collect a large amount of comic portraits of real people drawn by a famous comic author.

When a color image $x_a$ in $X(a)$ and an original grey image $y_a$ in $Y(a)$ are input into CycleGAN, $G_a$ uses $x_a$ as input and $y_a$ as a training teacher to output comic portraits, $fake_a$. Next, $G_b$ uses $fake_a$ as input and $x_a$ as a training teacher to output a refined image $rec_a$. Then the cycle-consistency loss, which is used to preserve the features of an original image, is determined by calculating the difference between $rec_a$ and $x_a$. This cycle-consistency loss will then be added to entire loss of GAN model. The above procedures are the first half cycle of CycleGAN.

In the second half cycle, CycleGAN executes $G_b$ first, by using $y_a$ as input and $x_a$ as training teacher to generate $fake_b$. Next, $G_a$ uses $fake_b$ to generate $rec_b$. Again, the cycle-consistency loss is determined by calculating the difference between $rec_b$ and $y_a$, and added to entire loss of GAN model (Fig. 3).

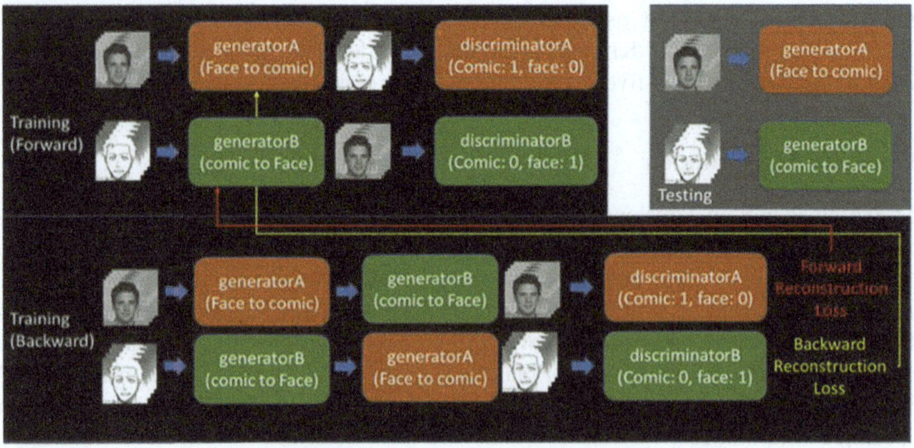


**Fig. 3.** Architecture of CycleGAN.

Although the CycleGAN can synthesize comic styled portraits, it tends to cause overfitting easily and lose important features of photos with noticeable noises (Fig. 4). Therefore, another GAN, PIX2PIX-based GAN is utilized based on the observations that synthesized facial texture can preserve the important features of real faces.

When training PIX2PIX-based GAN, input images are manually segmented into face parts and hair parts in the manner of divide-and-conquer (Fig. 5). In addition, the stroke lines are blurred to make the strokes more clear (Fig. 6).

**Fig. 4.** Result of CycleGAN.

**Fig. 5.** Colored comic pictures are separated into face parts and hair parts.

**Fig. 6.** Stroke lines are blurred.

# 4 Experimental Results

## 4.1 First Dataset

This section shows examples of synthesizing comic portraits using the proposed system, in which three comic author's painting styles were used for training. The first test dataset is a mixture of Kubo Taito's artwork, "BLEACH". In general, stroke lines of nose and mouth are simplified in most comics. Nose parts were often drawn as the shadow under it, and mouth parts were often ignored. However, in BLEACH, many stroke lines and screen tones are used as visual effects in the facial parts of characters.

As shown in Fig. 7, each photo is converted into the image below it, in which the stroke lines in mouth parts and nose parts were drawn clearly. It was also confirmed that the stroke lines of faces were similar to the shape of real faces, but with more emphasized manner in the result of morphing. It was also found there were some effects like screen tones in the result. In addition, the hair part in the result were drawn by separated lines. These results show the synthesized images reflect the comic style of BLEACH.

**Fig. 7.** Experimental results (using characters in "BLEACH" as resource).

### 4.2 Second Dataset

The second example aims to synthesize comic portraits using a paint style of Ito Junji's comic works. Compared to BLEACH, his characters tend to be more realistic, with softer stroke lines and fewer screen tones regions in face or hair parts. There are lots of greyscales and ink-filled regions in his comic works, which have to be separated by colors. For example, there are some ink-filled regions in hair areas, and some shadow areas filled with pencil strokes. In his comic works, some of the facial parts such as the bridge of nose are ignored (Fig. 8).

**Fig. 8.** Experimental results (using characters in Mr. Ito Junji's comic works as resource).

As shown in Fig. 8, shaded and shadowed areas in real photos were retouched with pencil strokes. In addition, hair areas were filled with ink effects. The stroke lines of these characters were softer, and the bridge of their noses were ignored. The results show the synthesized images reflect the comic styles of Ito Junji.

### 4.3 Third Dataset

The third example aims to synthesize comic portraits using a paint style of Sorachi Hideaki's comic work "Gintama." Compared to other two examples, the stroke lines of these characters are more abstractive and more simplified. Therefore, the difference between these character and real person are bigger. There are fewer greyscales and ink-filled regions then other comics. On the other hand, there are more scenes with exaggerated style, for expressing the feeling or personality of characters (Fig. 9).

**Fig. 9.** Experimental results (using characters in "Gintama" as resource).

As shown in Fig. 9, there were few screen tones regions in the result, and some of the stroke lines were ignored. Also, shaded and shadowed areas were replaced with few lines. Furthermore, eyes were drawn more clearly than other facial parts: using only few lines could still express the feeling of characters clearly. The results show the synthesized images reflect the comic styles of Gintama.

## 5 Conclusion and Future Work

This paper proposed a GAN system for synthesizing comic portraits. The proposed system combines CycleGAN and PIX2PIX-based GAN to reduce the effect of overfitting, which causes the result image collapse. The proposed system preserves most of the outlines of the input images and converts them to the comic strokes. Real face photos are therefore successfully transformed into vivid comic portraits with the painting styles of realism. Experimental results show that the system can synthesizes comic portraits based on three famous comic styles.

As for future work, it is necessary to improve the quality of the synthesized images. Because elements such as color tones, shadow or accessory on the characters may cause problems in the morphing phase, we plan to separate training process into several phases to improve the performance. For example, the first phase morphs eye parts only and nose parts are morphed in the subsequent process. This is expected to raise the quality of morphing for each part. It is also important to collect more comics from other countries or comic portraits drawn by professional comic authors for training to increase the success rates of morphing. In addition, shape exaggeration is an important and difficult challenge for comic arts. During the training of CycleGAN, some photos (less than 10% in CycleGAN) could be transformed into comic portraits with abstract and exaggerative effect (Fig. 10). Because finding the best value for training data may cause overfitting and make the result filled with noise, replacing stroke lines of facial parts with coordinate system by using displacement of the coordinate system could solve this problem. It is expected that displacement of the coordinate system could change the facial part's shape and location directly. After above-mentioned improvement, the proposed system is going to be evaluated qualitatively with experiments by recruiting comic fans and users of social software like Facebook and Line.

**Fig. 10.** Face photos that can be transformed into comic portraits with effect of shape exaggeration.

## References

1. Ledig, C., et al.: Photo-realistic single image super-resolution using a generative adversarial network. In: Conference on Computer Vision and Pattern Recognition (2017)

2. Zhu, J., Park, T., Isola, P., Efros, A.A.: Unpaired image-to-image translation using cycle-consistent adversarial networks. In: International Conference on Computer Vision (2017)
3. Ronneberger, O., Fischer, P., Brox, T.: U-Net: convolutional networks for biomedical image segmentation. In: Navab, N., Hornegger, J., Wells, W.M., Frangi, A.F. (eds.) MICCAI 2015. LNCS, vol. 9351, pp. 234–241. Springer, Cham (2015). https://doi.org/10.1007/978-3-319-24574-4_28
4. Isola, P., Zhu, J., Zhou, T., Efros, A.A.: Image-to-image translation with conditional adversarial networks. In: Conference on Computer Vision and Pattern Recognition (2016)
5. Zhang, H., et al.: StackGAN: text to photo-realistic image synthesis with stacked generative adversarial networks. In: International Conference on Computer Vision (2017)
6. Zhang, H., et al.: StackGAN++: realistic image synthesis with stacked generative adversarial networks. Inst. Electr. Electron. Eng. Trans. Pattern Anal. Mach. Intell. **41**, 1947-1962 (2018)
7. Radford, A., Metz, L., Chintala, S.: Unsupervised representation learning with deep convolutional generative adversarial networks. In: International Conference on Learning Representations (2016)
8. Chen, L., Chen, Y., Lin, S., Liu, T., Hsieh, W.: Synthesizing non-photorealistic rendering effects of volumetric strokes. J. Inf. Sci. Eng. **28**, 521–535 (2012)
9. Sun, C., Chen, L., Takama, Y.: Synthesizing NPR styled street view animation based on deep learning. In: International Conference on Technologies and Applications of Artificial Intelligence (2019)
10. Chen, S., Su, W., Gao, L., Xia, S., Fu, H.: Deep generation of face images from sketches. In: Special Interest Group on Computer Graphics and Interactive Techniques (2020)
11. Karras, T., Laine, S., Ail, T.: A style-based generator architecture for generative adversarial networks. In: Conference on Computer Vision and Pattern Recognition (2019)
12. Long, J., Shelhamer, E., Darrell, T.: Fully convolutional networks for semantic segmentation. In: Conference on Computer Vision and Pattern Recognition (2015)
13. Demir, U., Unal, G.: Patch-based image inpainting with generative adversarial networks. arXiv preprint, arXiv:1803.07422 (2018)
14. Goodfellow, I.J., et al.: Image augmentation using radial transform for training deep neural network. In: Institute of Electrical and Electronics Engineers International Conference on Acoustics, Speech, and Signal Processing (2014)
15. Perez, L., Wang, J.: The Effectiveness of data augmentation in image classification using deep learning. arXiv preprint, arXiv:1712.04621 (2017)
16. Bloice, M.D., Stocker, C., Holzinger, A.: Augmentor: an image augmentation library for machine learning. J. Open Sour. Softw. (2017)

# Road Surface Translation Under Snow-Covered and Semantic Segmentation for Snow Hazard Index

Takato Yasuno(✉), Hiroaki Sugawara, and Junichiro Fujii

Research Institute for Infrastructure Paradigm Shift, Yachiyo Engineering Co., Ltd., 5-20-8, Asakusabashi, Taito-ku, Tokyo 111-8648, Japan
{tk-yasuno,sugawara,jn-fujii}@yachiyo-eng.co.jp

**Abstract.** In 2020, there was a record heavy snowfall owing to climate change. In reality, 2,000 vehicles were stuck on the highway for three days. Because of the freezing of the road surface, 10 vehicles had a billiard accident. Road managers are required to provide indicators to alert drivers regarding snow cover at hazardous locations. This study proposes a deep learning application with live image post-processing to automatically calculate a snow hazard ratio indicator. First, the road surface hidden under snow is translated using a generative adversarial network, pix2pix. Second, snow-covered and road surface classes are detected by semantic segmentation using DeepLabv3+ with MobileNet as a backbone. Based on these trained networks, we automatically compute the road to snow rate hazard index, indicating the amount of snow covered on the road surface. We demonstrate the applied results to 1,155 live snow images of the cold region in Japan. We mention the usefulness and the practical robustness of our study.

**Keywords:** Scene understanding · Generative deep learning · Image-to-image translation · Semantic segmentation · Snow hazard index

## 1 Introduction

### 1.1 Background and Related Works

In Japan, approximately 60% of land is legally designated as the snow area and cold area, and approximately 20% of people have inhabited regions such as the Hokkaido, North Tohoku, Hokuriku, and Sanin. In the past two decades, approximately 4,000 traffic accidents have occurred per year during winter [1]. The primary reason includes slippery road (91%), reduced ability to see (6%), and rutted road (2%). In winter, following road conditions triggered accidents: iced (50%), dried (29%), wet (11%), and snow (10%). There are several primary hazards of winter driving [2]. They are: 1) poor traction, 2) reduced ability to stop vehicles, 3) the effect of temperature on starting and stopping vehicles, 4) slippery road conditions owing to ice and snow, 5) reduced ability to see and be seen, and 6) jackknifing. Recently, on January 10, 2021, owing to heavy snowfall on the Hokuriku Expressway, approximately 1,000 cars were stuck on the up and down lines


Y. Takama et al. (Eds.): JSAI 2021, AISC 1423, pp. 81–93, 2022.
https://doi.org/10.1007/978-3-030-96451-1_8

between the Fukui and Kanazu interchange. This congestion was caused by a slippery accident. Therefore, road surface monitoring is essential for road safety.

In the past decade, the focus has always been on an intelligent system for monitoring winter road surface conditions via computer vision and machine learning [3–14]. Takahashi et al. studied a prediction method for winter road surfaces corresponding to the change in snow removal levels [4]. Further, the road surface temperature and condition were estimated based on field observations. However, this method required several observed weather inputs such as temperature, wind velocity, humidity, sky radiation, and sunshine and incurred a higher cost for utilizing multi-mode data. Grabowski et al. proposed an intelligent road sign system including a detector using convolutional neural networks (CNNs) and image processing video cameras for the classification of dry, wet, and snowy roads [13]. Liang et al. investigated the application of semantic segmentation network customized with dilated convolutions for road surface status recognition [14]. The focus has been on performance improvement for road surface recognition; however, an important post-process of providing hazardous information on the winter road surface condition for road managers and users still remains.

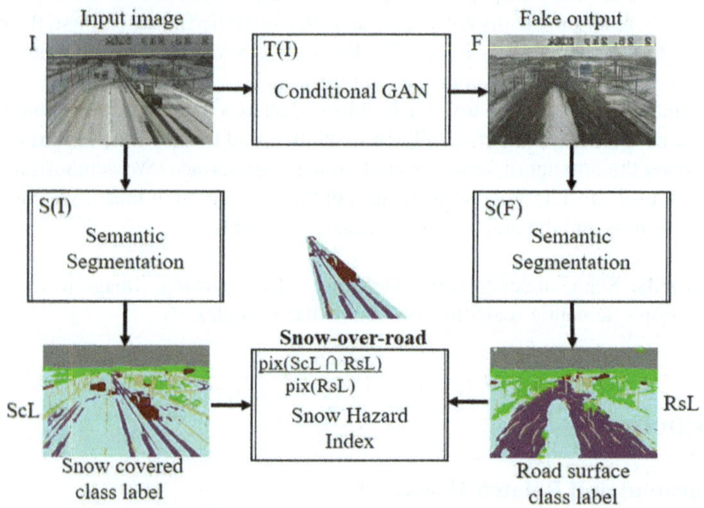


**Fig. 1.** Overview of our pipeline showing the raw input is translated by the conditional GAN network T. The resulting road surface fake output F is further fed into the semantic segmentation network S to predict a *road surface class label* output RsL. Simultaneously, the raw input is predicted into a *snow covered class label* output ScL. Finally, the intersection of both the outputs is the dangerous region of interest (ROI) *snow-over-road*. The ratio index based on two counting pixels of ROIs is computed by dividing the snow-over-road ROI by the road surface class ROI.

## 1.2 Monitoring Road Surface for Snow Hazard Alert

As illustrated in Fig. 1, we propose an application to provide snow hazard index information with snow hazard alerts using live camera images. If drivers can acquire information on snow levels through live cameras before their travel, they can decide to use snow adaptive tires to avoid tripping, in case of wet and slippery road conditions. The increasing usage of CCTV cameras has opened the possibility of using them to automatically detect hazardous road surfaces and inform drivers through alerts. In winter, we aim to provide a new snow hazard index on the amount of snow covered on the road surface, whose value is from zero to 100. During a heavy snowfall, the road surface region is covered with snow and is not clearly visible on the live camera. Through live image, the background snow excluded with road region would not influence road user for snow hazard. Thus, we need to know the ratio of snow covered on the road.

This study proposes a deep learning application with live image post-processing to automatically compute a snow hazard indicator. First, the road surface region hidden under snow is translated using a conditional generative adversarial network (GAN). Second, snow-covered and road surface classes are detected using semantic segmentation. Further, the road to snow rate hazard index is automatically computed, indicating the amount of snow covered on the road surface. Finally, the pipeline are applied on the snow and cold regions in Japan.

## 2 Deep Learning Application

Our proposed application contains the conditional GAN, semantic segmentation, and image processing. We formulate each task as follows.

### 2.1 Conditional GAN for Generating Road Surface Region

The image-to-image translation is possible for training a paired image dataset with fixed camera angle. This study proposes a method to generate the road surface from snow-covered region using L1-Conditional GAN. In the original pix2pix study, the input edge images were translated to shoe images [15]. Further, using the Yosemite photo dataset, they translated the summer photo to winter photo. However, in the case of snow-covered road images, it is difficult to detect the road surface region. This study applies a generative deep learning method, in which the road surface region hidden under snow is translated using the conditional GAN, pix2pix.

### 2.2 Semantic Segmentation for Snow and Road Classes

There are useful architectures such as semantic segmentation algorithms using deep learning for image analysis. The first end-to-end semantic segmentation neural network is the fully convolutional network (FCN) [16]. Several modifications have been made to FCN to improve performance. SegNet refines the encoder-decoder network layers with skip connections and conv-batchnorm-relu layers [17]. DeepLab replaced the deconvolutional layers with densely convolution and atrous spatial pyramid pooling instead of standard convolution layers [18]. This study applies a semantic segmentation method to detect snow-covered and road surface classes using a DeepLabv3+ algorithm with MobileNet as a backbone. This network has low memory and is practically accurate and fast for live camera monitoring of winter roads.

### 2.3 Post Image Processing for Snow Hazard Index

The input raw image is abbreviated as I. Based on the aforementioned deep learning, two trained networks were used: a conditional GAN network T and a semantic segmentation network S. As depicted in Fig. 2, we propose a pipeline to automatically compute a snow hazard index of the amount of snow covered on the road surface. First, a raw input image is translated into a road surface fake output F = T(I) using the trained pix2pix network T. In addition, the road surface output F is indirectly detected into the *road surface class label* output RsL using the trained semantic segmentation network S. The first image process is represented as two mapping function.

$$T(I) \rightarrow \mathrm{F}, \quad S(F) = S(T(I)) \rightarrow RsL. \tag{1}$$

Thus, we are able to focus on the monitoring scope ROI, which is the road surface region, though it has been hidden on the raw snow-covered input image. Notably, only the raw input image enable to approximate the road surface scope. Simultaneously, a raw input image is directly detected into the *snow covered class label* output ScL = S(I) using the trained semantic segmentation network S. The second image process is expressed as a next mapping.

$$S(I) \rightarrow \mathrm{Sc}L. \tag{2}$$

Furthermore, the intersection of both the outputs is highlighted on the dangerous target ROI *snow-over-road*. The ratio index based on two counting pixels of ROIs is computed by dividing the snow-over-road ROI by the road surface class ROI. Thus, a *snow hazard ratio* index, SHR, is formulated as next equations by counting pixels.

$$SHR = pix(RsL \cap ScL)/pix(RsL). \tag{3}$$

Here, pix(L) represents the pixel count of the class label region L. This snow hazard ratio index is represented as the function of the only input image I, conditional to the trained networks such as the conditional GAN translator network T and semantic segmentation network S.

$$SHR(I|T, S) = pix(S(T(I)) \cap S(I))/pix(S(T(I))). \tag{4}$$

Thus, the snow hazard ratio index is automatically computed using only the input raw image, instead of using the ground-truth image of the road surface without snow.

## 3 Applied Results

Herein, we demonstrate that our pipeline could automatically compute a snow hazard ratio index at the cold region. Figure 2 at the left side depicts the 16 test images randomly selected. In contrast, Fig. 2 at the right side illustrates the ground-truth images of the road surface without snow. Figure 3 depicts a road surface fake output translated from a raw input using a trained pix2pix.

### 3.1 Generating Road Surface from Snow-Covered Regions

We prepared 1,155 one-to-one datasets that included two groups: the snow covered road and the road surface without snow. We collected live road images from December 2 2020 to January 21 2021 at the cold region in Japan [19]. We unified 480 × 720-size images for training. The ratio of training data versus test data was 90:10. We set the input size of pix2pix network as 512 × 768. We iterated 200 epoch using the Adam optimizer; therefore, it took 65 h.

**Fig. 2.** Input raw images (left) and Ground-truth road surface (right) examples from [19].

**Fig. 3.** Fake output road surface generated from input raw images using pix2pix.

### 3.2 Prediction of Snow and Road Surface Segments

We prepared 500 annotated datasets that contained the raw images and six class defined labels such as road, pole-sign, green, snow, sky, and background. We resized images of 598 × 1196 size and extracted 2 × 4 crops, each with size 299 × 299; therefore, the total number of datasets was 4,000. The ratio of training data versus test data was 93:7. We set the input size of a semantic segmentation as 299 × 299. We trained 30 epoch iterations with a mini-batch of 16 and learning rate of 0.001 using the Adam optimizer;

therefore, it took 3.5 h. As presented in Fig. 4, each class mIoU had almost high accuracy scores, with the snow and road classes having the same score of more than 0.8. The entire accuracy mIoU was 0.7718, mean accuracy was 0.8572, and mean F1 score was 0.6886.

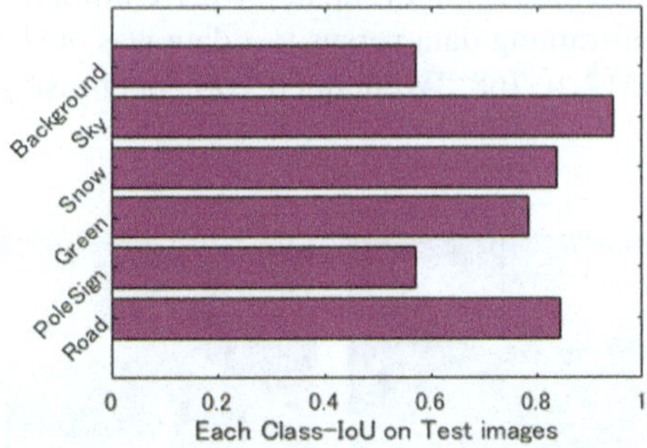


**Fig. 4.** Each class mean intersection of union (mIoU) results

As displayed in Fig. 5, the road surface fake output was segmented into the road class label output using the trained network DeepLabv3+ with MobileNet as a backbone. As depicted in Fig. 6, the input image was segmented into the snow class label output using the trained deep network.

**Fig. 5.** Road surface label results predicted from the road surface fake outputs using DeepLabv3+.

### 3.3 Computing Snow Hazard Index for Test Images

Figure 7 presents the pixel count for 16 image results for the *snow-over-road* (light cyan) and the *road without snow* (purple) ROIs. Each image has different size of road surface ROI. Each snow region does not include the background snow, which never influences the safety of the road user. Figure 8 presents the applied results of *the snow hazard ratio index*. Thus, it could automatically provide road managers and users with the alert information of snow over the road.

**Fig. 6.** Snow region label results predicted from the input raw images using DeepLabv3+.

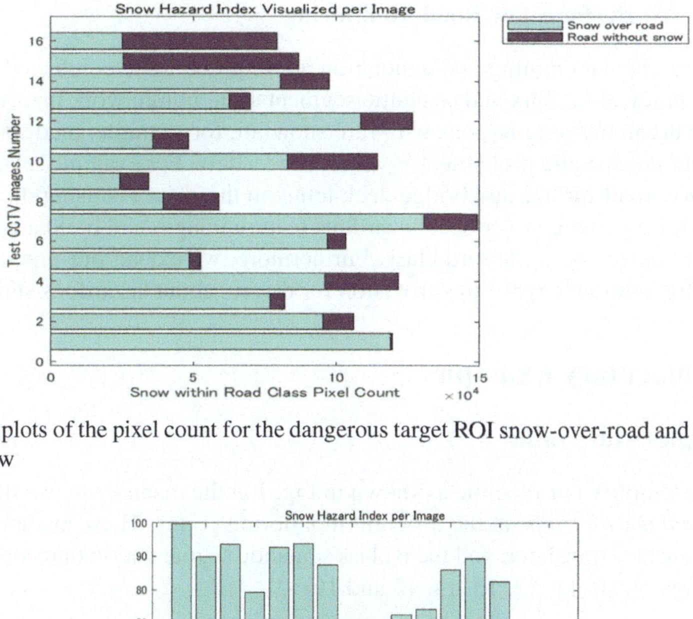


**Fig. 7.** Bar plots of the pixel count for the dangerous target ROI snow-over-road and road surface without snow

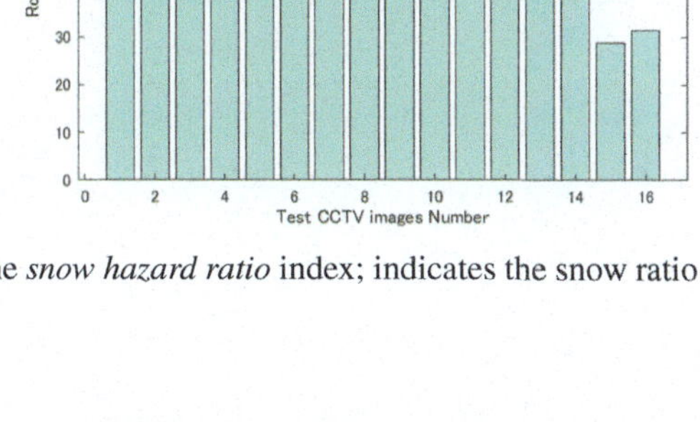


**Fig. 8.** Bar plots of the *snow hazard ratio* index; indicates the snow ratio within the road surface ROI of each image.

# 4 Concluding Remarks

## 4.1 One Image Translator and Segmentation Application for Snow Hazard Index

This study proposed a deep learning application with live image to automatically compute a snow hazard ratio indicator. First, the road surface hidden under snow was translated using a conditional GAN, pix2pix. Second, hazardous target ROI snow-over-road and road surface region were segmented using DeepLabv3+ with MobileNet as a backbone. Further, the snow hazard ratio index was computed for the amount of snow covered on the road surface. We demonstrated our pipeline to the cold region in Japan. This application had the advantage of using only one live image as the input, without any mining of before and after paired images dataset. Furthermore, the indicator could be delivered to the road managers and users using our pipeline automatically computed *snow hazard ratio index*, whose value was zero to 100 for multi-points comparison.

## 4.2 Future Works for Snow Road Monitoring

The In future, the data mining and annotation are required to be conducted to achieve a higher accuracy of pix2pix and semantic segmentation. Future work involves transfer learning for urban traffic in regions with rare snowfall, for example, metropolitan area. Further, there are specific problems for winter road safety. For example, black ice on a seemingly wet road surface and bridge deck icing on the river. Focusing on the critical snow hazard, we continue to collect hazardous feature images and build a classifier for preprocess to detect strong hazard class. Furthermore, we extend our pipeline to edge computing for automatic real-time provision for drivers about hazardous snow points.

# 5 Supplementary Examples

## 5.1 Examples with Traffic

In order to exemplify our pipeline as shown in Fig. 1 at the manuscript, we demonstrate 16 *test examples with traffic* at the different monitoring points. These are applied based on the road surface translator, and the 6 class semantic segmentation that contains snow and road (Figs. 9, 10, 11, 12, 13, 14, 15 and 16).

**Input and Ground-truth Road Surface, Generated Road Surface Snow-covered**

**Fig. 9.** Input raw images examples From [19] (left), and fake output (right) road surface generated from input raw images using pix2pix.

**Predicted Snow and Road Surface Segmentation**

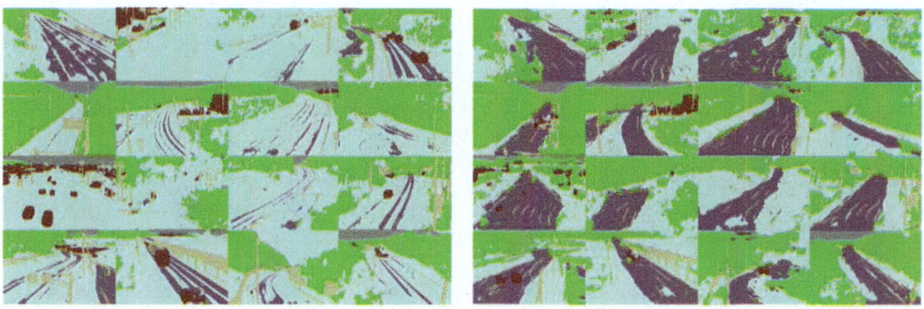

**Fig. 10.** Snow region label predicted results (left) from the input raw images and Road surface label predicted results (right) from the road surface fake outputs using DeepLabv3+.

## 5.2 Multi-period Examples

Furthermore, in order to exemplify our pipeline as shown in Fig. 1 at the manuscript, we demonstrate the four camera inputs of *temporal variants* with four multi-period test examples. These were collected at 2021 January 19 to 21 in this winter. These are applied based on the road surface translator, and the 6 class semantic segmentation including snow and road.

**Input Road Surface and Generated Road Surface from Snow-covered**

Fig. 11. Input raw images (left) examples from [19] and Fake output (right) road surface generated from input raw images using pix2pix.

**Predicted Snow and Road Surface Segmentation**

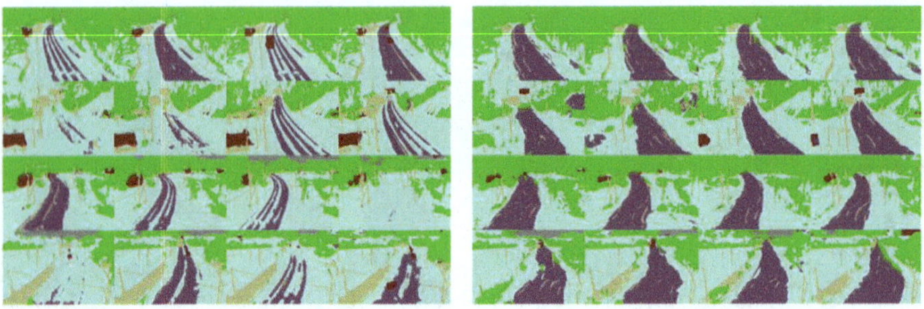

Fig. 12. Snow region label predicted results (left) from the input raw images and Road surface label predicted results (right) from road surface fake outputs using DeepLabv3+.

## 5.3 Snow Hazard Index Output Examples

Finally, in order to exemplify our indicator as shown in Fig. 1 at the manuscript, we demonstrate the snow hazard index, which are computed from two previous examples such as the examples with traffic, and multi-period examples.

## Computed Hazard Index for Images with Traffic

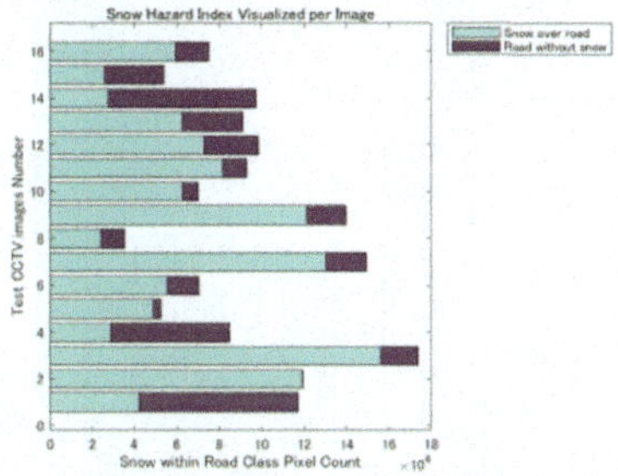


**Fig. 13.** Examples with traffic, bar plot of the counting pixels for the dangerous target ROI snow-over-road and road surface without snow.

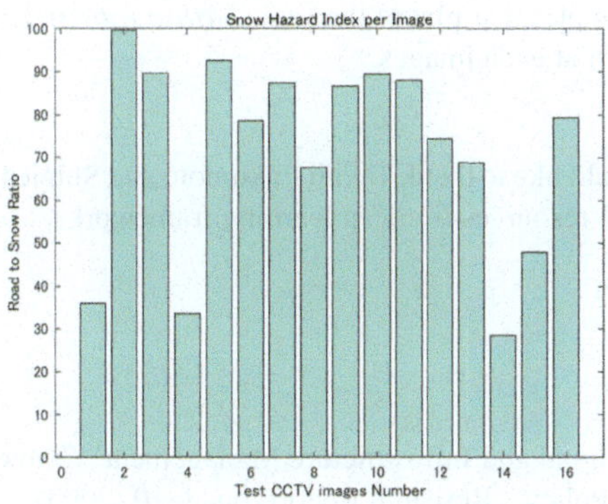


**Fig. 14.** Examples with traffic, bar plot of the *snow hazard ratio* index, indicates the snow ratio within the road surface ROI at each images.

## Computed Hazard Index for for Multi-period Images

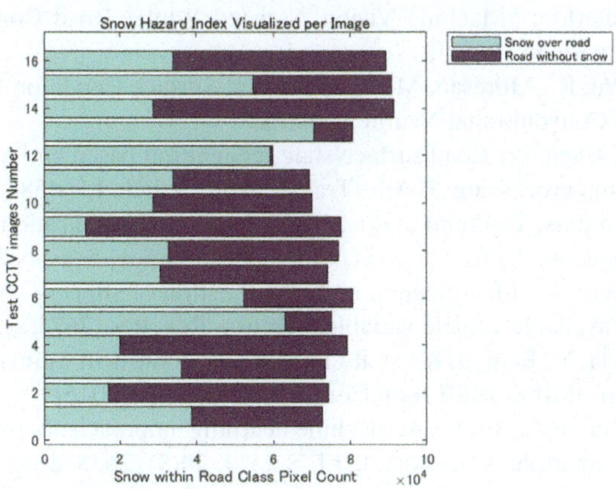


**Fig. 15.** Multi-period examples, bar plot of the counting pixels for the dangerous target ROI snow-over-road and road surface without snow.

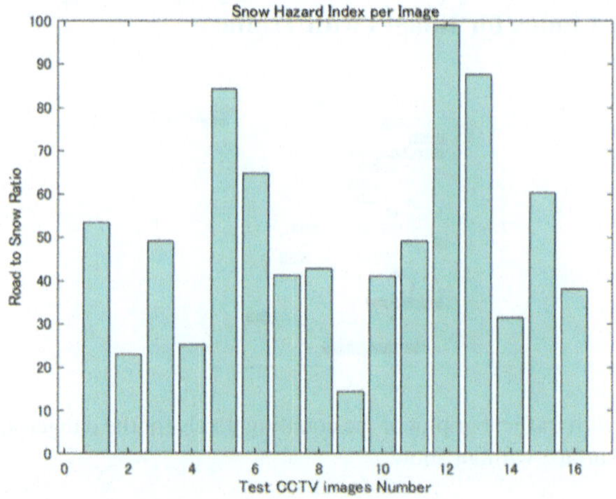


**Fig. 16.** Multi-period examples, bar plot of the *snow hazard ratio* index, indicates the snow ratio within the road surface ROI at each images.

**Acknowledgments.** I would like to thank Takuji Fukumoto and Shinichi Kuramoto (MathWorks) for providing us MATLAB resources for deep learning frameworks.

## References

1. National Institute for Land and Infrastructure Management, "Snow and Cold Region Traffic Accident Survey and Policy", Research Report, pp. 3–10 (2003)
2. Saskatchewan Professional Driver's Handbook, pp. 136–139. Saskatchewan Publication (2019)
3. Loce, R.P., Bernal, E.A., Wu, W., Bala, R.: Computer vision in roadway transportation system: a survey. J. Electr. Imaging **22**(4), 041121 (2013)
4. Takahashi, T., Tokunaga, R., Sato, K., et al.: A Study on Development of Prediction Method for Winter Road Surface Corresponding to Change of Snow Removal Level (2012–2015)
5. University of Waterloo: "Machine Vision Assisted Winter Road Condition Monitoring", Technology Summary, 8810-7336
6. Pan, G., Fu, L., Yu, R., Muresan, M.: Winter Road Surface Condition Recognition Using a Pre-trained Deep Convolutional Neural Netwrok
7. Zhao, J., Wu, H., Chen, L.: Road surface state recognition based on SVM optimization and image segmentation processing. J. Adv. Transp. **2017**, Article Id 6458495, Hindawi (2017)
8. Sabanovic, E., Zuraulis, V., Prentkovskis, O., Skrickij, V.: Identification of road surface type using deep neural networks for friction coefficient estimation. Sensors **2020**, 20,612 (2020)
9. Padarthy, M., Heyns, E.: Identification and classification of slippery winter road conditions using commonly available vehicle variables. Transp. Res. Rec. **2673**(2), 60–70 (2019)
10. Nagata, Y., Kaneda, Y., Echigo, K., et al.: Potential of snow drift estimation by using CCTV camera images. In: JSSI & JSSE Joint Conference, Nagoya (2016)
11. Basavaraju, J., Du, F.Z., Ji, J.: A machine learning approach to road surface anomaly assessment using smartphone sensors. IEEE Sens. J. **20**(5), 2635–2647 (2020)

12. Carrillo, J., Crowley, M.: Integration of roadside camera images and weather data for monitoring winter road surface conditions. In: Canadian Association of Road Safety Professionals CARSP Conference, (2019)
13. Grabowski, D., Czyzewski, A.: System for monitoring road slippery based on CCTV cameras and convolutional neural networks. J. Intell. Inf. Syst. **55**, 521–534 (2020)
14. Liang, C., Ge, J., Zhang, W., Gui, K., Cheikh, F.A., Ye, L.: Winter road surface status recognition using deep semantic segmentation network. In: International Workshop on Atmospheric Icing Structures (2019)
15. Isola, P., et al.: Image-to-image translation with conditional adversarial network. In: CVPR (2017)
16. Shelhamer, E., Long, J., Darrell, T.: Fully convolutional networks for semantic segmentation. IEEE Trans. Pattern Anal. Mach. Intell. **39**, 640–651 (2017)
17. Badrinarayanan, V., Kendall, A., Cipolla, R.: SegNet: a deep convolutional encoder-decoder architecture for image segmentation. IEEE Trans. Pattern Anal. Mach. Intell. **39**, 2481–2495 (2017)
18. Chen, L.C., Papandreou, G., Kokkinos, I., et al.: DeepLab: semantic image segmentation with deep convolutional nets, atrous convolution, and fully connected CRFs. IEEE Trans. Pattern Anal. Mach. Intell. **40**, 834–848 (2018)
19. Ministry of Land, Infrastructure, Transport and Tourism, the live camera images of each region office activities in Japan. https://www.mlit.go.jp/road/road_e/index_e.html
20. Gonzalez, R.C., Woods, R.E.: Digital Image Processing, 3rd edn. Prentice Hall (2008)
21. Foster, D.: Generative Deep Learning: Teaching Machines to Paint, Write, Compose and Play. O'reilly (2019)
22. Yasuno, T., Fujii, J., Amakata, M.: Pop-outs segmentation for concrete prognosis indices using UAV monitoring and dense dilated convolutions. In: International Workshop on Structural Health Monitoring, Stanford, CA (2019)

# Generation of Infographics Considering Information Recipients

Shota Miyoshi, Yuji Nozaki, and Maki Sakamoto(✉)

Graduate School of Informatics and Engineering, The University of Electro-Communications, Chofugaoka 1-5-1, Chofu, Tokyo 1828585, Japan
maki.sakamoto@uec.ac.jp

**Abstract.** Infographics convey information in attractive and effective way by combining various types of data, such as texts and pictures. Although variety of infographics designing tools have been proposed, most of them generally require proficient skill and deep understanding of the software upon operation. In this paper, we propose a user-friendly infographic generator by using natural language as its input. In addition, considering that the same information can bring different feelings to people by changing a writing style of texts and numerical expressions, our system provides a function to change the base color of infographic so that user can control the impression of infographic as "positive" or "negative". For evaluation of quality of infographics, we conducted a questionnaire survey comparing infographics of different colors that assumed to affect an impression of presented information. As the results of analysis, it was found that the color-controlled infographics are more efficient to enhance the impression in the intended direction than comparative images without color control.

**Keywords:** Infographics · Natural language processing · Emotional color wheel

## 1 Introduction

It is generally known that information with visual elements, such as pictures, chart, and graphs, is easier to understand than that without them [1–3]. In today's information-oriented society where the situation changes rapidly, it became more and more important to convey information quickly and correctly. Thus, techniques for visualization of variety kinds of information are important for this purpose [4,5]. There are various methods for visualizing information [6,7]. For example, as a familiar visualization method, bar graphs and line graphs are often used to express the distribution and changes of numerical values.

An infographic is one of the methods of visualizing information with numerical values to aid peoples' understanding using both text and images [8,9]. As infographics can also make the content more attractive, they are used in many fields such as news, signage, materials of conference, and marketing [10,11].


Y. Takama et al. (Eds.): JSAI 2021, AISC 1423, pp. 94–105, 2022.
https://doi.org/10.1007/978-3-030-96451-1_9

A huge demand of infographics exists in various fields of news reporting including SNS as well.

To lighten a workload for creating infographics, various kinds of methods and tools have been proposed. For example, some presentation software has a function that helps layout arrangement, change in sizing, and transformation of input images. In this research, we developed a system that drew infographics include pieces of fundamental information, such as subject words and numerical expressions that automatically extracted from given texts by natural language processing. For this purpose, our system performs series of conventional natural language processing techniques. Specifically, it performs sentence division, parsing, labeling onto the given text and then identify the fundamental information based on multiple rules we defined. In visualization, a pictogram which matches the most for the given text is selected from the prepared dataset by searching with meta tags and is used in the generated infographic. Also, as a means of controlling the emotions given to the recipient of information, the color of the visual element is automatically changed by inputting two polarities, positive or negative, designated by the user of the system.

As described, the proposed method does not require any knowledge or operation skills of the software. This advantage allows everyone to use infographics with ease. Also our method offers a function of modifying the color of infographics considering impression that you want to give to the recipient. Thus, our system is expected to be useful in the fields requiring appealing of contents, such as news and marketing in SNS.

## 2 Related Researches

### 2.1 Infographics

Infographics are visual representations of information, data, and knowledge [12]. Infographics can be used to convey information attractively through visual effects [13]. Infographics has existed since a long time ago, and various research related to generating Infographics have been reported. For example, research on DDG (Data-Driven Guides), a data-driven design tool by Kim [14]. In DDG, there is a concept called a "data guide". Here, the main visual variables of length and area are first determined, and they are represented as linear and circular guides. Then, the guide is converted to the final output infographic through familiar operations as image editing such as rotation, enlargement, reduction, and moving, and binding processing with data.

While DDG is a tool for design experts, there is research on data-driven design tools that focus on ease of use for inexperienced designers. InfoNice [15] is a tool that adds infographic design functionality to Microsoft Power BI[1]. Microsoft Power BI is a service that creates basic graphs such as line graphs and pie charts from numerical information stored in the database, and InfoNice provides design functions by inserting infographic conversion processing in the

[1] https://powerbi.microsoft.com/.

output results of Power BI. Although creating infographics with InfoNice is easier than with DDG, there are restrictions on the layout.

A method for generating infographics from sentences has also been proposed by Cui [16]. In this method, there are four visual space variables namely layout, description, graphic, and color. These were introduced with reference to Power Point including texts related to infographics such as "n%" and "m-in-n" under designer's supervision. The explanatory sentences adopted in Infographics are the entire sentence of the input sentence, the sentence excluding the numerical value, and/or a part of the verb or object. Seven types of graphs, which are frequently used in infographics, are used as graphics. The color is determined using the pre-installed color scheme tools. The layout, which the display position of graphics and explanations, is determined by using prepared template. The final infographic output is generated by applying the determined explanations and graphics to the template. As described in this section, assistant tools for drawing infographics generally require a user proficient skill and deep understanding upon operation. In this paper, we propose a user-friendly assistance system for creating infographics by using natural language as its input.

### 2.2 Natural Language Processing

In general, infographics are required to concisely display major nouns and numbers. With our methods introduced in this paper, nouns and numerical values are automatically extracted from the text input to the system. Named entities and quantities contained in text are called *named entities*, and there is an extended named entity hierarchy defined to classify Japanese named entity into several classes [17]. General named entities include personal names, organization names, place names, and also product names and facility names. There are three large classes in the extended named entity hierarchy called *name*, *time*, and *numerical.* The *name* includes a person's name and a place name, and the *time* expression includes a time and a date. *Numerical* expression includes amount, ratio, frequency, etc. The extended named entity hierarchy is created to meet the needs of a wide range of named entity in information extraction, information retrieval, automated question answering systems and machine translation, and is expected to be used in a wide range of natural language processing. Note that our original methods only work for Japanese sentence, thus all examples appear in this papers are manually translated into English for the purpose of explanation.

## 3 Implementation

Our system consists of two parts: text analysis processing and infographic creation processing. The working flow to creating an infographic from a given sentence is as follows: (1) In the text analysis process, the information used for the infographic is extracted from the input information. The input information given to the system is Japanese text and negative or positive emotional polarity. (2) In order to simplify the information extraction process, the input text targeted by our system is a text that contains up to two numerical information. (3) In the statistic name and statistic value are determined by analyzing the input text with Japanese parsing software. The statistic name is the subject of the input text, and the statistic value is a number that is presumed to be related to the subject. (4) Then, the icon image (pictogram) and color of the visual element are determined based on the statistic name and polarity information. This visual element is picked up from the icon dataset in which the icon image and the color are linked. (5) Finally, the layout containing the visual elements that can correspond to the determined statistic is determined. The final output infographic is generated by applying the statistic, statistic name, icon image, and color to this layout.

### 3.1 Text Analysis

The text input made by the user is analyzed using a morphological analysis tool that supports Japanese[2,3]. For the numerical expressions that handled by this system, the following are used as expressions that are easy to compare and frequency: *N_Person, N_Organization, N_Location, N_Location_Other, N_Country, N_Facility, N_Product, N_Event, N_Natural_Object, N_Natural, _Object, _Other, N_Animal, N_Flora, Money, Percent, Multiplication*, and *Frequency*.

### 3.2 Selection of Subject Word

The system firstly parses the input text to determine the subject which to include in the output infographics. In this process, first, a word that can be the subject is searched from all the words contained in the sentence. Candidate words for the subject are tokens that have a dependency relationship with *nsubj*, which means the subject noun, or *obj*, which means the object. At this time, if the token that is the candidate of the subject is *obj*, we check whether the following two conditions are satisfied: if *obj* token does not include the numerical expression, or if an *nmod*, which means noun modifier, token appears before the *obj* token in the sentence. If both conditions are satisfied, this *obj* token is used as a candidate for the subject. This is because when *obj* token is a numerical expression in the sentence syntax, it is natural to regard that another *nmod* word, which thought to be the subject, has already appeared in the text.

If the input sentence includes multiple subject candidates, the subject candidates are filtered to be only one candidate that are most suitable as statistic name. This filtering process is performed considering the word location in the sentence and the relationship between the subject of the input text and the numerical expression. First, a distance between the subject and the token labeled with one of the numerical expressions is obtained by calculation as the number of tokens that exist between the two target tokens. If the subject is a one letter word, it is regarded as not a valid candidate as the subject. Therefore, in the

[2] https://spacy.io/.
[3] https://github.com/megagonlabs/ginza.

case of a one-letter subject, the distance between the subject and the numerical expression was penalized to be 1.5 times larger. After the average value of the distance between the subject and each numerical expression is calculated, the one with the smallest average distance, that is, the word closest to multiple numerical expressions is adopted as the subject. Note that this rule is set taking characteristics of Japanese language into account. In Japanese language, one letter subjects usually stand for large scale concepts rather than concrete subjects. For instance, *person, money*, and *number* are usually expressed with a one letter (=人, 金, 数) while *neighbor, charge*, and *mathematics* are two letter words (=隣人, 料金, 数学).

If the input text does not contain any subject, an alternative subject is chosen from the multiple keywords obtained from the input text. Here, each keyword is a group of nouns and words that describe them, such as proper noun tokens with the *obj* tag. Then, the number of nouns and other part of speech included in each keyword are counted, and the one with the most noun tag words among the keywords is adopted as the final subject. When there are multiple keywords containing the same number of nouns, the keyword with the smallest number of parts of speech other than nouns is adopted as the subject. If the subject is one character and the modifier of the subject is also one character, our system assumes that the one placed at the early part of the sentence has a greater role as the subject, in this case the proper noun of more than two letters, which appears before the adopted modifier, is chosen as the subject.

### 3.3 Selection of Values

This section describes how to extract the statistic to embed in Infographics from a given text. If the input text contains only one numerical representation, this one is used as the statistic of the input text. When the input text includes multiple numerical expressions, the relevance between the subject and each numerical expression is calculated, except for "Percent" and "Multiplication" which mean percentages. First, if the word with the numerical expression tag and the subject has a dependency relationship or a modification relationship, the degree of association between the subject and the corresponding numerical expression is incremented by 1. If the numerical representation token is the subject itself, the degree of association is not incremented. Finally, the numerical expression token with the highest degree of association is selected as a candidate for numerical expression in the input text. If there are tokens with the same type of numerical representation as the candidate numerical representation, those tokens are used as the numerical representation in the input text. For example, with the given text "The price of new product is changed from 200 yen to 220 yen for 3 pieces", the *amount* is "200 yen" and "220 yen", and the *number of products* expression is given to "3 pieces". When "new product" is chosen as the subject, the degree of association of "220 yen" with "new product" is higher than the degree of association of "3 pieces" and "200 yen". In this case, therefore, "220 yen", which has the highest degree of relevance to the subject, and "200 yen", which is the same type of numerical expression as this numerical expression, are used as the final numerical expressions, and "3 pieces" is excluded from the numerical expressions to be used in the infographic. Words labeled with *percent* and *multiplication* are regarded as changes in the selected numerical representation. *Percent* or *multiplication* is only selected as the numerical representation when any other numerical representation is not included in the input text. Our system is designed for input texts containing only one of *Percent* or *Multiplication* at most.

### 3.4 Change in Values

The infographic generated by our system contains information about the changes in the numbers contained in the given text. Therefore, we created a customized dictionary focusing on changes in numerical values. For example, if the word "exceeds" pertains to a statistic, it implies the statistic may rise. On the other hand, if it is "below", it implies the statistic decreases.

To create the dictionary, we manually labeled words related to numerical changes based on publicly available securities reports and web news articles. Words that imply the statistic goes up are labeled *UP*, and words that imply the statistic goes down are labeled *DOWN*. As the result, 26 words were registered for each *UP* and *DOWN* label. This numerical change dictionary is connected to the language analysis tool so that the numerical change label is given in the same way that the input text of this system is given the numerical expression label by the analysis tool.

The basic rules for determining numerical fluctuations are as follows: (1) if the numerical change label is attached to the word related to the subject, the numerical change label attached to the word is regarded as the numerical variation in the input text. (2) if the word with the numerical change label is related to the word related to the subject, the numerical change label given to the word related to the word is regarded as the numerical variation in the input text. (3) If a word with a numerical change label or a word that is dependent on the word is related to a numerical expression, the numerical change label given to the word is regarded as a numerical change. (4) If multiple numerical change labels are included, the numerical change label at the end of the sentence is used as the numerical change in the input text.

If the input text does not contain any words with a numeric change label, or if the *Percent* or *Multiplication* token contains *UP*, or *DOWN* words. Numerical fluctuations are determined based on this label classification. For example, "20% increase" includes two tokens, "20%" and "increase", the numerical fluctuation of the input text is *UP*. Similarly, if the token of *"N_xxx"* (for example, *N_Person*) contains *UP* or *DOWN* words, the numerical fluctuation is determined based on this label classification. If a word with a numerical change follows after a word with a token of *Percent* or *Multiplication*, the value converted from the words is used in the output graph. In this conversion, for example, "10% decrease" is converted to "90%", "30% increase" is converted to "130%", and "1.5 times" is converted to "150%". Statistics can also be normalized based on numerical fluctuations. With this manner, from a sentence containing "10% reduction", the converted value of "90%" will be appeared in a graph of infographic.

### 3.5 Icon

We prepared a dataset for icon images to be displayed in the infographic. These icons are selected from the dataset based on the statistic name obtained by text processing, First, a data set containing all the icon image data used in this system was created. In this dataset, one icon image is associated with multiple tags and one color. For the icon, we use the image listed in ICOON MONO[4], where related tags are attached to these icon images. For example, the icon image representing spinach has tags such as "vegetables" in addition to "spinach". As the result, we obtained a dataset contains a total of 6,432 types of icon images and a total of 8,878 tags.

Since the original icons are a monochrome image, we determined the representative color associated with each icon by following procedure: first, we performed a Google image search using the collected icon image names as a key, and obtained the top 10 images in the search results. For the RGB representation of these 10 images, K-Means clustering was performed to extract dominant colors, and the color most frequently used was adopted as the representative color of the icon. Considering that the representative color is biased toward black and white, we apply a custom rule of $r > 200$ *and* $g > 200$ *and* $b > 200$ and $r < 55$ *and* $g < 55$ *and* $b < 55$ to RGB value of dominant color candidates so that a color with low saturation was not adopted as the representative color.

Then, we determined one icon image that most related to the obtained statistic name from the icon dataset. In this system, the degree of relevance between the tag and the statistic name was determined by calculating the word distribution expression included in the icon dataset. fastText was used for word vectorization [18]. The corpus used for learning Japanese word relevance was "jawiki 20201120" [19], and we used NEologd [20] for morphological analysis dictionary. The tags to be searched in the data set were restricted to the tags that have a vector representation in the fastText model (6,314/8,887 words). The word distribution expression of the statistic name and the tag is acquired by fastText, and the tag with the largest cosine similarity is used as the icon related to the statistic name.

### 3.6 Color

This chapter describes the algorithm to determine the colors used in the infographic. In this system, the colors are determined based on the representative color associated with the icon in the dataset and the negative or positive polarity values given by a user. We use the HSB (H: hue, S: saturation, B: brightness) color system for the color model of our system. First, we assign corresponding hue ranges for each of the positive and negative. For the mapping of hue regions, we referred to the "wheel of emotions" proposed by Plutchik [21]. As a result, the range corresponding to the positive was set to the range of 0 to 135, which is the hue corresponding to "expectation" and "joy", and the range corresponding

[4] https://icooon-mono.com/.

to the negative was set to the range of 136 to 360 corresponding to "sadness", "disgust", etc.

The color obtained from the representative color associated with the icon (hereinafter referred to as Color A) is mapped on the HSB color wheel (hereinafter referred to as Color A'). Based on this A', a plurality of color candidates of different hues is selected. The color candidates are: (1) two colors (Color B1, B2) that differ by 120 °C from A', and (2) three colors (Color C1, C2, C3) that differ by 90 °C from A'.

Then, considering the input polarity, one of the six colors (A', B1, B2, C1, C2, C3) that falls into the positive or negative range that determined earlier is selected for the further steps (hereinafter referred to as Color D). To prevent the colors used in the infographic from becoming grayscale, color D has been adjusted the lower limit of saturation of color D is 50 and the lower limit of lightness of color D is 60, while preserving its hue (Color D1'). Finally, the colors (D'2, D'3) whose hue is ±30 based on D1' are acquired, and D'1, D'2, and D'3 are used as the colors for the infographic to be generated.

As an example, let us follow our algorithm the case where the color A associated with the icon is #6aaa30 (H: 91, S: 72, B: 67). First, B1 = (211, 72, 67), B2 = (331, 72, 67) are obtained as the hue is ±120 from A', and C1 = (181, 72, 67), C2 = (271, 72, 67), C3 = (1, 72, 67) are obtained as the hue is ±90 from A'. If the input polarity information is positive, one of A' = (91, 72, 67) and C3 = (1, 72, 67) whose hue is between 0 and 135, is randomly selected. Here, let us assume that C3 is selected this time (= D'1). Finally, D'1= (1, 72, 67), D'2 = (31, 72, 67), and D'3 = (331, 72, 67) with a hue of ±30 from D'1 are obtained for the infographics.

### 3.7 Layout

A layout template for infographic is selected from five prepared templates. Each template has several boxes to embed extracted information as shown in Figs. 1 and 2. As the number of extracted information varies depending on the input text, the template that has the same number of boxes will be chosen. Figure 3 shows examples of infographic using templates.

**Fig. 1.** Layout templates

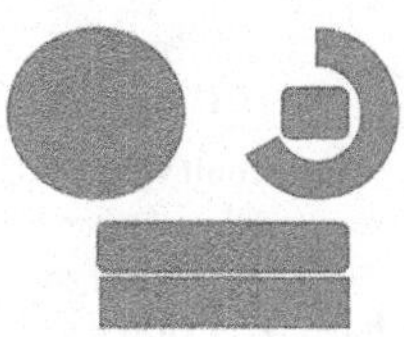

**Fig. 2.** Layout templates

(a) The harvest of tomato descreased 20% from last year.

(b) The farewell party was held with 30 people.

(c) The gas price increased by 1.2 times from 133 yen to 158 yen.

**Fig. 3.** Examples of infographic

## 4 Evaluation

We conducted a questionnaire survey for comparison of infographics with different colors that assumed to affect an impression of information (Fig. 4). 32 (24 males and 8 females, aged between 16 and 57, mean age = 24.6) people were participated in the survey. To create variety of different infographics, we collected 26 texts from web news pages and then modified their subject and numerical expressions. Each text was presented to the subjects with the infographic generated by our system with different emotion color. Each participant is asked to answer the following two questions on a scale of 1 to 7 considering a view point of buyer or seller. This condition was repeatedly told to participant on each question, and each participant was asked to answer only one designated condition(buyer or seller). Other explanations on infographics were not made during the experiment. The subjects were divided into two groups. The subjects belong the Group A were asked to answer the questionnaires from the standpoint of a seller of products. In contrast, the subjects in Group B were asked to answer as a buyer of products.

1. What impression do you get from infographics? (1 = Negative, 7 = Positive)
2. Do you think the infographic matches with the text? (1 = No, 7 = Yes)

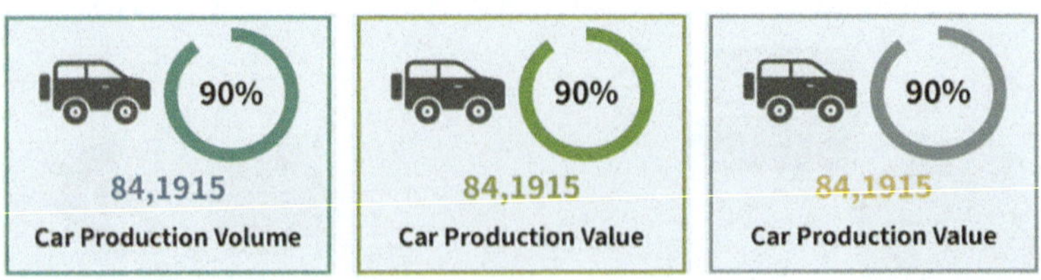


**Fig. 4.** Polarity(Negative - Positive) and comparative image

To compare difference in impression, infographics created with the dominant colors from image search results was used as a comparison image. Table 1 shows the result of questionnaire 1. In case the input polarity information is negative, the result is counted as "true" if the mean score of the emotional polarity image is smaller than the mean score of the comparison image. Likewise, in case the input polarity information is positive, the result is "true" if the mean score of the emotional polarity image is larger than the mean score of the comparison image. We defined this rule as a "normal" comparison, and the comparison with and is defined as a "strict" comparison. Finally, we calculate the mean value by dividing the number of "true" by the number of images.

**Table 1.** Results of questionnaire 1 in each classification

| | Seller | Buyer | Negative | Positive |
|---|---|---|---|---|
| Normal comparison | 0.62 | 0.65 | 0.62 | 0.65 |
| Strict comparison | 0.54 | 0.46 | 0.38 | 0.62 |

In the normal comparison, more than 60% of the results ware "true". Thus, it can be said that proposed image was efficient to give the intended impression to participants. On the other hand, the result of the strict comparison became lower score. This trend is exhibited clearly in the "negative" classification scores (approximately 38%). We have investigated the cases where the result of normal comparison rule is true but the result of strict comparison rule was false and found that the comparable images used for the corresponding question tended to evaluated as non-positive. In other words, the comparative image used for the question that we expect to get a "negative" result gave a strong negative impression to subjects as the original state. This fact can be considered to be one of the factors that affect the evaluation of the negative strict comparison.

**Table 2.** The mean score (standard deviation) of evaluation of emotional polarity image and comparative image, and results of one-side t-tests

| | Polarity modified image | Comparative image | P-value |
|---|---|---|---|
| Seller | 3.35 (2.03) | 3.12 (1.99) | 0.25 |
| Buyer | 3.89 (2.25) | 3.88 (2.14) | 0.95 |
| Negative | 3.49 (2.11) | 3.20 (2.03) | 0.16 |
| Positive | 3.75 (2.21) | 3.80 (2.12) | 0.82 |
| Average | 3.62 (2.16) | 3.50 (2.10) | 0.42 |

Table 2 shows the results of questionnaire 2. As shown in the Table 2, there were no statistically significant differences between polarity modified images and comparative images. However, it can be seen that the average values of answers of two different standpoints tend to lead to more negative in polar images than in comparative images for both sellers and buyers. From the viewpoint of considering the validity of the polarity information used for image generation, we compared the average value of each condition. As the result, subjects are more likely to feel infographics matches to text with the positive controlled images, but less likely with the negative control images. In summary, the polar-controlled images whose hues were manipulated by our system were more likely to cause information mismatches to the subjects than the non-manipulated images.

## 5 Discussion and Conclusion

In this paper, we proposed a method for generating an infographic from a given text by extracting fundamental information such as subjects and numerical values with natural language processing. The greatest advantage of our method is that infographics can be generated without requiring the user to have complicated knowledge or experience. This advantage allows everyone to use infographics with ease. In addition, our method offers automatic color modification function. The basic color of Infographics is the color obtained from image search, and the color can be modified to the "positive/negative" direction whichever the user wants to give to the recipient of information. As the result of the analysis on questionnaire survey, it was found that the color-controlled infographics are more efficient to enhance the impression to the recipient in the intended direction than comparative infographics without color control. These results are consistent with past studies in the field of color psychology [21].

On the other hand, we found some shortcomings as the result of evaluation, such as relatively lower quality of matchingness between text and infographics. This may be due to a problem with our color scheme algorithm or extracted information was insufficient for recipients. Regarding the color scheme, we used very simple HSB color wheel base algorithm in our system, however, using more sophisticated color modification algorithms those provided in commercial software widely used by designers (for example, Adobe color palette [22]) may improve the quality of infographics to be generated. Introducing a color scheme that takes into account the saturation contrast, area contrast, or the relationship between shape and color may also contribute to improvement in quality of infographics [23]. In the proposed method, we considered two directions for changing the impression: negative and positive. However, emotions of human are greatly complicated, thus more emotional expression may useful as some SNS offer 4 or more emotional reactions, e.g. like, sad, angry, etc. It would be more convenient for users if the generator automatically selects a color scheme and layout that suits these emotions with other advanced machine learning techniques.

## References

1. Mayer, R.E., Gallini, J.K.: When is an illustration worth ten thousand words? J. Educ. Psychol. **82**, 715–726 (1990)

2. Borkin, M.A., et al.: What makes a visualization memorable? IEEE Trans. Vis. Comput. Graph. **19**(12), 2306–2315 (2013)
3. Bateman, S., Mandryk, R.L., Gutwin, C., Genest, A., McDine, D., Brooks, C.: Useful junk? The effects of visual embellishment on comprehension and memorability of charts. In: Proceedings of the SIGCHI Conference on Human Factors in Computing Systems, CHI 2010, pp. 2573–2582 (2010)
4. Hullman, J., Adar, E., Shah, P.: Benefitting InfoVis with visual difficulties. IEEE Trans. Vis. Comput. Graph. **17**, 2213–2222 (2011)
5. Levie, W.H., Lentz, R.: Effects of text illustrations: a review of research. Educ. Commun. Technol. **30**(4), 195–232 (1982)
6. Heer, J., Bostock, M., Ogievetsky, V.: A tour through the visualization zoo. Commun. ACM **53**(6), 59–67 (2010)
7. Cairo, A.: The Functional Art: An Introduction to Information Graphics and Visualization. New Riders Press (2012)
8. Harrison, L., Reinecke, K., Chang, R.: Infographic aesthetics: designing for the first impression. In: Proceedings of the 33rd Annual ACM Conference on Human Factors in Computing Systems, CHI 2015, pp. 1187–1190 (2015)
9. Dunlap, J.C., Lownthal, P.R.: Getting graphic about infographics: design lessons learned from popular infographics. J. Vis. Literacy **35**, 42–59 (2016)
10. Dur, B.İ.U.: Interactive infographics on the internet. Online J. Art Des. **2**(4), 1–14 (2014)
11. Huang, S., et al.: The effect of an infographic promotion on research dissemination and readership: a randomized controlled trial. Can. J. Emerg. Med. **20**, 826–833 (2018)
12. Jun, S.: Basic Infographics. BNN, Inc. (2013)
13. Haroz, S., Kosara, R., Franconeri, S.L.: ISOTYPE visualization: working memory, performance, and engagement with pictographs. In: Proceedings of the 33rd Annual ACM Conference on Human Factors in Computing Systems, CHI 2015, pp. 1191–1200 (2015)
14. Kim, N.W., et al.: Data-driven guides: supporting expressive design for information graphics. IEEE Trans. Vis. Comput. Graph. **23**(1), 491–500 (2017)
15. Wang, Y., et al.: InfoNice: easy creation of information graphics. In: Proceedings of the 2018 CHI Conference on Human Factors in Computing Systems, CHI 2018, no. 335, pp. 1–12 (2018)
16. Cui, W., et al.: Text-to-Viz: automatic generation of infographics from proportion-related natural language statements. IEEE Trans. Vis. Comput. Graph. **26**, 906–916 (2020)
17. Extended Named Entities - Ver 8.0.0. http://liat-aip.sakura.ne.jp/ene/ene8/definition_jp/html/home.html
18. Bojanowski, P., Grave, E., Joulin, A., Mikolov, T.: Enriching word vectors with subword information. Trans. Assoc. Comput. Linguist. **5**, 135–146 (2017)
19. Wikimedia Downloads. https://dumps.wikimedia.org/
20. mecab-ipadic-NEologd : Neologism dictionary for MeCab. https://github.com/neologd/mecab-ipadic-neologd
21. Plutchik, R.: A general psychoevolutionary theory of emotion. In: Theories of Emotion, pp. 3–33 (1980)
22. Adobe Color. https://color.adobe.com/
23. Itten, J., Ohchi, H.: The Elements of Color, vol. 12, pp. 33–62. Bijutsu Shuppan-Sha, Ltd. (1989)

# The Best Place in Tokyo for Anything: Classifying Stations by Walkability to Specific Amenities

Aaron Bramson[1,2,3(✉)] and Megumi Hori[1]

[1] GA Technologies, Inc., Roppongi 3-2-1, Minato-ku, Tokyo 106-6290, Japan
a_bramson@ga-tech.co.jp
[2] RIKEN, Center for Biosystems Dynamics Research, Minatojima-Minamimachi 6-7-3, Chuo-ku, Kobe 650-0047, Japan
[3] Department of General Economics, Ghent University, Tweekerkenstraat 2, 9000 Ghent, Belgium

**Abstract.** Walkability analyses have recently gained attention for economic, health, and environmental reasons [5,8,13]. We examine the walkable store profiles for train stations of central Tokyo and use machine learning to find clusters of similarly walkable areas. First, we use a breadth-first search algorithm on the road network to determine the walkable areas within 5, 10, and 15 min of each station. We then collect the establishments within 50 m of any traversed edge. We perform three analyses: (1) classifying regions by the numbers of stores of each type, (2) recursive feature selection and reclassification, and (3) scoring areas by their specialization in one of the store categories. We find that classification without feature selection produces more useful results, and that the <15 min isochrones yield the best results. These methods can be broadened to identify regions that are over- and under-serviced by amenities with impacts for both policy and business planning.

**Keywords:** Walkability · Geospatial analysis · Machine learning · Network analysis · Transportation networks

## 1 Introduction

With increasing concerns over the environmental, economic, and health effects of an overly car-dependent society, research into walkability has gain significant momentum. The end-goal is to find ways to improve the walkability of areas, but there are still many conceptual and methodological issues needing to be resolved. Walkability is not a single measure, but rather a subclass of accessibility measures. Accessibility measures describe the degree to which specific amenities (healthy food, day care, sports facilities, etc.) can be reached within an area or from some point, and 'walkability' measures the ability to reach those places by foot. The current work focuses on access to the full range of establishments generally related to everyday life.


Y. Takama et al. (Eds.): JSAI 2021, AISC 1423, pp. 106–117, 2022.
https://doi.org/10.1007/978-3-030-96451-1_10

We use machine learning to cluster areas by their store category profiles and identify areas that specialize in some specific category. We define regions as the walkable areas from each of the train stations within the 23 wards of Tokyo. We collect the stores within each walkable area to generate a store profile for each station. We use these profiles to cluster regions with similar numbers across store types. Then we perform feature selection on the profiles to determine which store categories have the greatest impact in differentiating regions' walkability. We follow this up with an analysis of the level of specialization of areas within each category; i.e., to what degree an area specializes in restaurants, cafes, supermarkets, etc. The analysis here is exploratory, demonstrating the use of AI to examine and group geographic areas by their characteristics, so we finish by describing the many avenues by which we can expand and apply this research going forward.

## 2 Data

Our analysis requires the integration of three disparate datasets: the walking network, train stations, and stores.

**Store Data.** Our store data comes from Open Street Map (OSM) [11]. For the current demonstration, we limit our analysis to establishments within selected categories relevant to walkability. We defined these categories from the fine-grained store types in OSM to group together similar stores rather than dealing with hundreds of specific store types. For example, the various stores that sell only one kind of food item (chocolates, cake, cheese, etc.) become grouped together as "specialty food". There are a total 56,493 establishments within central Tokyo, partitioned into the 25 mutually exclusive categories shown in Fig. 1 along with their counts. For simplicity we refer to all these establishments as "stores".

**Walking Network Data.** Our walking network data is the "road" network from Open Street Map (OSM) [11], which includes pedestrian-only paths, but excludes roads where walking is marked as prohibited (such as expressways). We also collect the nodes and edges for subway and train station entrances; if they are disconnected from the road network, we link them to the nearest road node. We measure the lengths of each edge using a distance-preserving projection centered on Tokyo Station, and the traversal time is calculated using a walking speed of 4.8 kph (80 m per minute).

**Train Station Data.** Because our analysis focuses on train stations, we incorporate the locations of railway, subway, and streetcar stations using [9] for the station and line names, and merge this with the station nodes from OSM [11] to get accurate locations. Stations for different lines that share the same name are captured as separate stations. We filtered 27 of the 600 stations that were not properly integrated into the surrounding network. Because entrance nodes in the OSM data typically do not specify for which station they are an entrance, we simply connect each entrance node to its nearest station node to integrate the

**Fig. 1.** Each of the selected categories of establishments used for calculating walkability along with its count with Tokyo's central 23 wards.

stations with the walking network. The resulting integrated network has 731,570 nodes and 893,187 edges.

## 3 Methods

We describe our method for integrating the stations into the walking network above; now we describe our methods for generating walkable areas, clustering areas by similar store profiles, feature selection, and identifying area specialities.

**Determining the Walkable Area.** Starting from each station we perform a single source Dijkstra algorithm using the NetworkX library [6] to calculate the walking times to each node within 15 min. We select three isochrones (<5, <10, <15 mins) to test the sensitivity of the results to the stores' distances from the station. For each isochrone, we take the set of edges traversed within the appropriate time frame and add a 50m buffer to its geometry (after converting its geometry into a distance-preserving coordinate reference system centered on Tokyo station). We then use the exterior of the resulting polygon in order to fill in holes left from the union of buffered edges. The buffers for all stations can be seen in Fig. 2. Note that because the network and store data are all isolated within central Tokyo, there is an edge effect for stations near the boundary. Figure 2 offers a detailed view for two sample stations. In this figure you can also see the high level of detail of our network and the extent of a 50m buffer applied to all the traversed edges.

**Clustering By Establishment Profiles.** Using the counts of stores in each category as the feature vector, we use SciKit-Learn's [12] hierarchical clustering with 7, 10, and 15 clusters to collect the stations into groups with similar store profiles. Before clustering, we standardize the data: for each category we find the

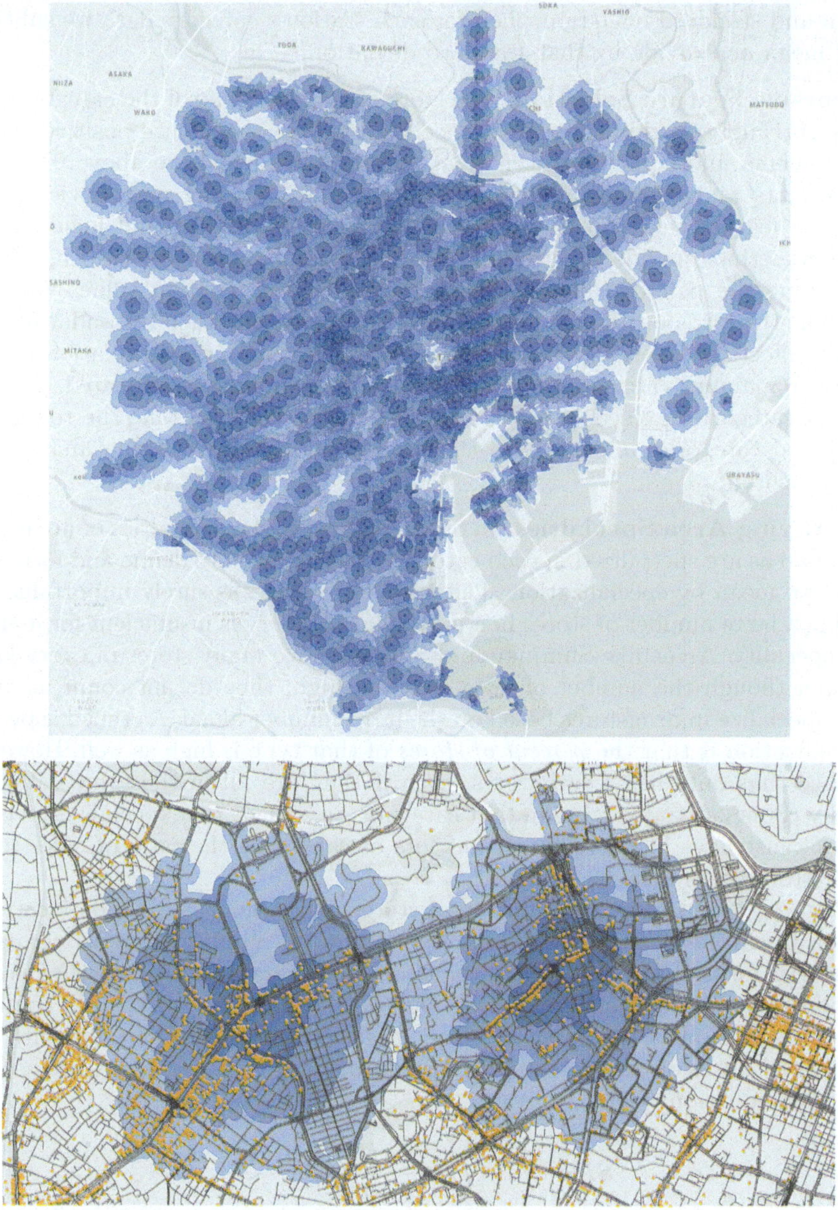


**Fig. 2.** (top) The (often overlapping) isochrones for 573 stations in the central Tokyo area. From the dark centerpoint marking the stations, the colors lighten in 5-min increments. (bottom) Detail of the walkable areas for two stations (Akasaka and Gaienmae) displaying three walkable area isochrones (<5, <10, <15 mins). Grey lines represent network edges and the gold dots mark store locations. [7, 10, 11]

mean and standard deviation, then for each station's category data we subtract that mean and divide by that standard deviation.

**Recursive Feature Selection.** The store counts for many of the categories are (unsurprisingly) highly correlated; specifically, all the value are positive except one (cinema and community center at −0.01) and several are above 0.90 (e.g. restaurants and cafes, gambling and bars, retail shopping and services). We perform feature selection to reduce the number of variables and focus on the categories that provide the greatest informational gain. For each isochrone, we determine the pair of categories with the highest correlation value. We then combine the raw store counts for those two categories and then restandardize the result, effectively merging the categories into one. Then, the correlation matrix of the remaining categories is recalculated. This process is repeated until no pair of categories exceeds a 0.50 correlation level. We then take the remaining combined categories and perform the same hierarchical clustering analysis for each isochrone.

**Identifying Area Specialties.** In order to identify the specialties of areas, and which areas are specialized in each category, we first need to define and formalize what we mean by specialization. The number of stores is surely important, but having a large number of stores in a particular category is insufficient for making it a specialty. Areas like Shinjuku and Shibuya have many stores of every kind, so even though the number of drug stores is high, they do not count as areas that specialize in drugstores based on the large number alone. A crucial aspect of specialization is that the *percent of stores* of that type is high as well. However, a place that only has a few stores, even if they are all in the same category, hardly counts as specializing in that category. So a place that specializes in cafes doesn't just have lots of cafes, it has lots of cafes *and* the ratio of cafes to all other store types is high.

Let $m_{c,i,b}$ represent the number of stores of category $c \in C$ for location $i$ within isochrone $b$, and let $N_{i,b}$ represent the total number of stores across all categories for that location within that isochrone. First we calculate the percent of stores for each category in the obvious way: $p_{c,i,b} = m_{c,i,b}/N_{i,b}$. To combine our measurement of the concentration of stores by category ($p_{c,i,b}$) with the number of stores of that category ($m_{c,i,b}$) into a specialization index, we use the following equation:

$$SI(c,i,b) = p_{c,i,b} \frac{\ln(1 + 5m_{c,i,b})}{1 + e^{(4-0.2m_{c,i,b})}} \tag{1}$$

The natural logarithm of the store counts captures diminishing returns on larger numbers of stores while maintaining that having more stores always yields greater specialization. In preliminary experiments we found that areas with small numbers and high percentages were given overly high SI scores. The sigmoidal component and constants in Eq. 1 are chosen manually to fit the data so that the two factors balance well across the observed range of count and percentage values.

## 4 Results

Overall, our analysis reveals intuitive groupings of stations by similarity depending on the isochrone distance and choice of variables in ways much to be expected. The specialization index, while still a rough measure of what we are interested in, does provide useful insights beyond what can be discerned from store count or percent values alone.

**Clustering on All Categories.** We performed tests exploring $\{7, 10, 15\}$ clusters, for brevity we focus on 15 clusters here. The maps in Fig. 3 display the clusters found for each isochrone. With the <15 min isochrones (bottom-right) we can see strong clustering around known city centers: central Shinjuku in light orange, surrounding Shinjuku plus Ikebukuro, Kasumigaseki, and Otemachi + Jimbocho in dark green, the Ginza + Shimbashi combined with the business area around Tokyo station in light green, and the Shibuya + Harajuku + Omotesando area in gray. For those familiar with the Tokyo area, these groupings are highly intuitive. Grouping nearby stations together is a minimal expectation because of how the walkable zones are created and the large isochrone areas. There is also shared groupings among distant stations among the less central locations. The purple, pink, and red stations scattered along commuter line stations capture variations in suburban store profiles.

Decreasing the walk time to <10 min maintains some of those clusters, but in general the identified groups become smaller and somewhat less identifiable. However, closer examination reveals that the <10 min isochrone (Fig. 3 bottom-left) performs well in revealing natural groupings. Grouping the areas around Shinjuku, Ikebukuro, and Shibuya together demonstrates that the store profile is capturing a genuine similarity among these famous shopping and entertainment areas. Ueno, Todai Hongo, and other, small areas near the center get their own groups. Also, Harajuku gets paired with non-adjacent Shimokitazawa, which was not observed in other experiments, but is a natural combination.

Comparing the <5 min isochrone (Fig. 3 top) we see that one station in the Ikebukuro area is in the same category as Hibiya, two others are grouped with major shopping areas in central Tokyo (light orange), and two others are in a group scattered around (brown). When it comes to very large stations, if the travel time is only 5 min, then which side of the main station you are on can have a big difference on the reachable stores. Such a high degree of sensitivity makes the <5 min results less reliable for capturing perceived levels of similar walkability.

**Clustering on Selected Features.** Correlation-based recursive feature selection identifies the store categories with the least redundant information with respect to finding groups with similar store profiles. Because the correlations differ by isochrone, we perform feature selection separately for each one. With increasing distance the number of features decreases from 23 to 2 for the <15 min isochrone, compared to 5 for the <10 min isochrone and 8 for the <5 min isochrone. The religion category is independent across all isochrones. Specifically, for <10 min we have religion, supermarket + laundry, community centers, and

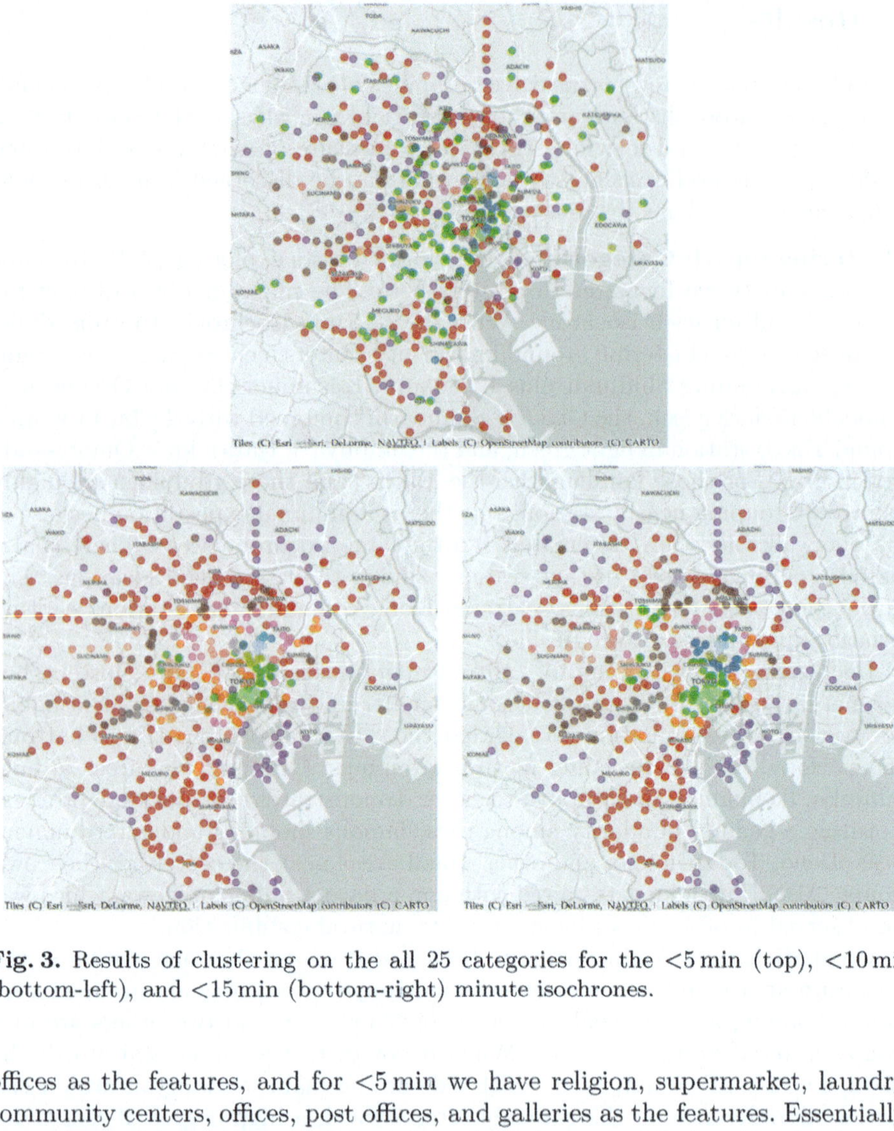

**Fig. 3.** Results of clustering on the all 25 categories for the <5 min (top), <10 min (bottom-left), and <15 min (bottom-right) minute isochrones.

offices as the features, and for <5 min we have religion, supermarket, laundry, community centers, offices, post offices, and galleries as the features. Essentially, for <15 min this has become the total number of stores while for other isochrones there is some remaining differentiation.

The clusters generated by the combined features (shown in Fig. 4) are markedly different than the clusters produced using all the variables, although there are some important consistencies as well. In both analyses, Ueno, Akihabara, and the area in between (Suehirocho and Okachimachi) each get their own distinct groups. However, many other clusters are split and recombined with other, not clearly related, stations. In the <15 min case, the dark green area that still includes Ikebukuro and the area surrounding Shinjuku, now also includes various commuter line hub stations. Although it is reasonable to think these

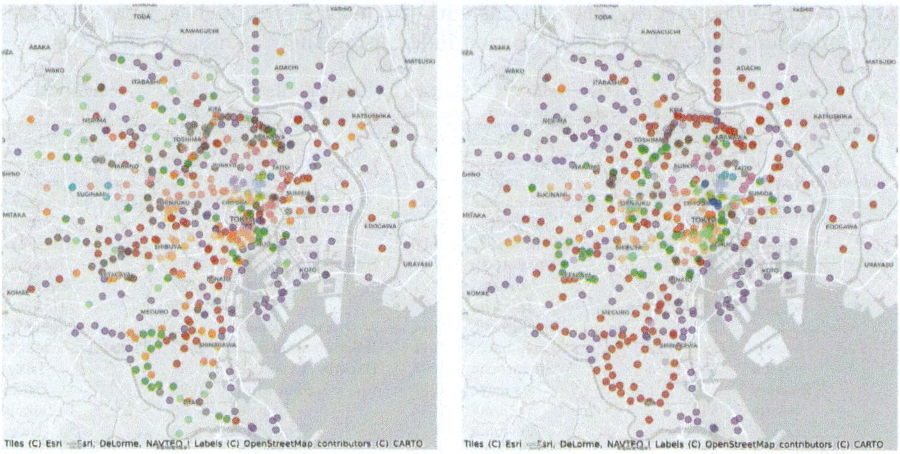

**Fig. 4.** Results of clustering on the combined categories from recursive feature selection for the <10 min (left) and <15 min (right) isochrones.

disparate areas have similar total store counts, they do not provide similar shopping or living experiences due to the differences in store profiles. The <10 min results are again more fragmented than the <15 min results, so although they carry more information about distinct profiles (2 → 5 groups), they may be overly sensitive to small variations in minor categories.

**Specialty Analysis.** Using Eq. 1 to combine the number and percentage of stores of each type we calculate the specialization index (SI) in terms of areas focused on that kind of store. We show the station with the largest number, the largest percent and largest specialization index along with those numbers for the <15 min isochrone in Table 1. From this table we can see most possible patterns among these variables: a higher concentration compensates for a lower number of stores (e.g. clinics in Kugahara vs Yushima), a large number suffices to outweigh a lower focus (e.g. community centers in Ogikubo vs Toneri-Koen), and the many cases where the balance of two is a different station than either of the ones with the greatest number or concentration.

Note the large differences in the specialization indices across categories. The highest category (restaurants) has a specialization index of 2.96 while the smallest two (cinemas and offices) have a specialization index of essentially 0.00. Partly this reflects the nature of different store types; the number of cinemas is just small compared to the numbers for other kinds of stores. So even an area that is relatively high in its number of cinemas is going to have too few of them to count as specialized in cinemas. This tells us that there really are no places that specialize in cinemas, galleries, fitness centers, gambling, etc. It also tells us that there are places that specialize in restaurants, shopping, convenience stores, and parks.

**Table 1.** For each category, the stations with the most stores, highest percent of local area stores, and the highest specialization index using the <15 min walkable area.

| category | largest number | station with largest number | highest percent | station with largest percent | highest SI | station with highest SI |
|---|---|---|---|---|---|---|
| restaurant | 1149 | Suehirocho | 0.45 | Shijō-mae | 2.96 | Nishi-shinjuku |
| shopping | 890 | Suehirocho | 0.26 | Shin-ochanomizu | 2.02 | Shin-ochanomizu |
| convenience store | 144 | Akihabara | 0.67 | Urayasu | 1.90 | Higashi-mukojima |
| park | 83 | Monzen-nakacho | 0.43 | Toneri-koen | 1.82 | Kiyosumi-shirakawa |
| bar | 840 | Suehirocho | 0.25 | Shinseibijo | 1.76 | Ueno |
| healthcare | 244 | Meiji-jingumae 'Harajuku' | 0.18 | Harajuku | 1.26 | Harajuku |
| cafe | 328 | Ochanomizu | 0.37 | Haneda Airport T2 | 1.02 | Harajuku |
| clinic | 201 | Yushima | 0.21 | Kugahara | 0.94 | Kugahara |
| religion | 113 | Inaricho | 0.27 | Toneri | 0.72 | Hakusan |
| specialty food | 129 | Suehirocho | 0.14 | Keisei Takasago | 0.66 | Sengoku |
| beauty | 158 | Okachimachi | 0.11 | Chikatetsu-akatsuka | 0.57 | Chikatetsu-akatsuka |
| bank | 75 | Suehirocho | 0.25 | Haneda Airport T3 | 0.44 | Tōkyō |
| laundry | 48 | Asagaya | 0.17 | Showajima | 0.36 | Ochiai |
| leisure | 83 | Yushima | 0.33 | Urayasu | 0.36 | Ikebukuro |
| spa | 169 | Yushima | 0.17 | Showajima | 0.33 | Nakai |
| supermarket | 37 | Suehirocho | 0.17 | Showajima | 0.16 | Nishi-nippori |
| service | 74 | Suehirocho | 0.21 | Ryutsu Center | 0.13 | Ochanomizu |
| post office | 23 | Shin-Nihombashi | 0.33 | Kasai-Rinkai Park | 0.07 | Ningyocho |
| community center | 18 | Ogikubo | 0.10 | Toneri-koen | 0.05 | Ogikubo |
| specialty school | 30 | Yushima | 0.04 | Kasai | 0.04 | Hongo-sanchome |
| gambling | 29 | Suehirocho | 0.03 | Kita-ayase | 0.03 | Ueno-okachimachi |
| fitness | 30 | Ueno-hirokoji | 0.05 | Shijō-mae | 0.03 | Ueno-hirokoji |
| gallery | 16 | Ginza-itchome | 0.02 | Tennozu Isle | 0.02 | Ginza-itchome |
| cinema | 6 | Yurakucho | 0.01 | Futako-shinchi | 0.00 | Yurakucho |
| office | 6 | Hongo-sanchome | 0.02 | Shimo-Igusa | 0.00 | Takadanobaba |

Instead of just looking at the locations with the maximum specialization values, we can also gain insight by looking at the geospatial distribution of stations with higher and lower scores. Figure 5 shows the specialization indices for all stations in central Tokyo for two related categories: restaurants (left) and bars (right). The plots tell us that business and shopping areas have both more and a higher percentage of eateries and bars than residential areas, and there are similarities in the patterns because of this. However, the distribution of bars places the Ueno area at the top, while Ikebukuro fails to indicate any specialization. Some of these patterns, however, may be artifacts of the data's categorizations. The line between restaurants and bars is fuzzy, especially when considering Japan's izakaya culture (a sort of pub that is frequently used for after business meetings). These results are not surprising; on the contrary, these results confirm that our specialization index is accurately tracking which stores are overly abundant in which areas.

That said, the specialization results do depend on the time horizon used. Changing isochrones often reveals the same or nearby stations as most highly specialized. However, changing the isochrone can have a large impact on stations that are near, but not too near, major commercial districts. For example, there are many retail shopping venues within 5 min of Suehirocho station, but within 15 min one can reach Ueno, Ochanomizu, and Akihabara and all the restaurants, bars, and cafes they contain. This larger scope dilutes the specialization of Suehirocho at larger isochrones.

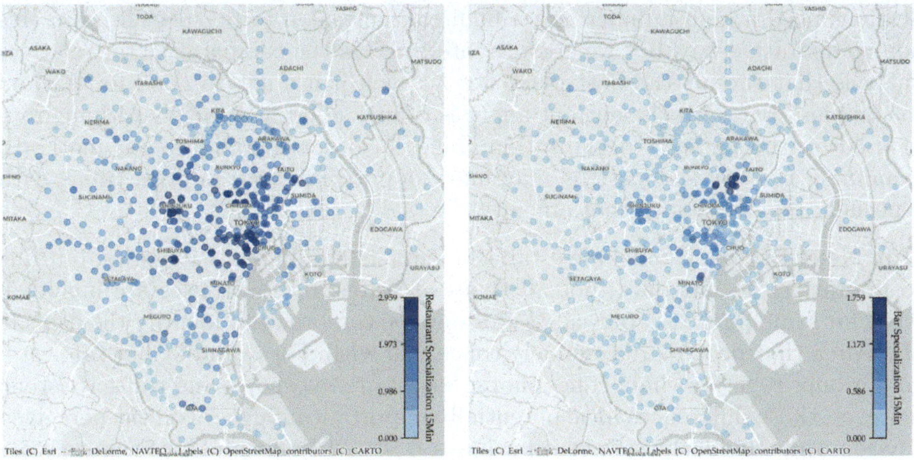


**Fig. 5.** Results of the speciality analysis for two categories: restaurants (left) and bars (right).

Because there is no ground truth for walkability scores or for specialization, the best we can say at this stage is that the specialization index captures and balances intuitively important aspects of specialization and that the results of the <10 min and <15 min isochrones largely match our experiences of the areas around Tokyo. The results also point to aspects that can be fruitfully improved, such as weighting the contribution of stores by their walking time [3], adjusting the equation parameters based on the total numbers by store type, and perhaps different aggregations of the hundreds of store types into categories.

## 5 Conclusions and Future Work

We found that hierarchical clustering with <10 min or <15 min isochrones reliably captures city centers and systemic patterns in peripheral areas in an intuitive way. Iterative feature selection by combining highly correlated categories simplifies the analysis by aggregating what should be largely redundant information, but the reduced categories lack the variety needed to capture differences in store profile types. At the extreme we find that for <15 min isochrones, nearly all categories are combined and the clustering results for the total numbers of stores fails to capture intuitively similar station areas. By clustering on the mix of stores we see the greatest concentration of groups in areas with the most stores (where differences in the numbers matter the most). If we want to see different groupings (such as urban vs suburban, entertainment vs business, dining out vs daily life) or based on differences in other features (e.g. transportation network features [2]) then we need to adjust the data used for clustering and perhaps use a more sophisticated clustering method.

The specialization analysis taught us that using a logistic log function allows us to create scores that balance the influence of store counts and percentages.

Stations need a high number and a high concentration to be specialized in this non-comparative way, but we are also interested in the comparative (or relative) analysis that reveals the best place for $X$, even if it's not so good for $X$, and to combine that with the current analysis of how good it is for $X$. Although we focused on the highest specialization values in order to validate our methods, examining the lowest locations may mark opportunities to improve areas with especially poor access to that amenity.

**Our Next Steps.** We include establishments in 25 different categories, but due to limitations in our data, not all types of stores and amenities are included. Also, for some purposes, our categories may be too broad to identify the aspects of an area that people are genuinely concerned with. For example, our cafe category groups together chain coffee shops, 'kissaten', tea shops, and coffee roasters; however, those differences have potentially important influences on the perceived environment. Similarly, for some purposes it would be helpful to separate store brands and/or premium establishments from ordinary ones (e.g. Natural Lawson from ordinary Lawson convenience stores). Along these lines, one may be interested in not merely the number of stores, but also on having a variety of each subcategory or brand within each category. We are limited by the accuracy and completeness of OSM data in this regard, but we plan to experiment with more sophisticated categorization algorithms.

In the current methodology we include stores that are within 50m of any road edge using three time horizons, and so some stores that should be reachable will not be included in the walkable area. Furthermore, using buffered networks we cannot weight the stores' contribution by walking time except within those three zones. Following [1] we could instead integrate the stores into the road network by creating nodes for them and connecting those nodes to the road network. This method would allow us to calculate the walking time to each store and use it to adjust the impact the store has on walkability.

Although focusing the analysis on scoring stations is natural in train-centric Tokyo [4], it does have its limitations. As apparent from the maps, there are gaps between stations that are not being sampled or scored. There is also significant overlap in the stations' isochrones, especially near the center, which yields stations with very similar results for large isochrones. Using a point grid as the origin of points can remove the gaps and irregular overlaps, but results in scores focused on arbitrary locations. Combining a fine-grained grid and time-weighted store counts may be the best solution for accurately and parsimoniously scoring and categorizing the walkability of localities.

As stated in the introduction, walkability scoring has gained popularity among researchers, city planners, and developers; however, research on walkability is still in its infancy. Aside from the variety of amenities that could be included or excluded from such a measure, there is also the question of how to capture and integrate those amenities for the analysis (buffered roads, distance weighting, integrated network, etc.). Clearly additional work is necessary to untangle these considerations and establish shared standards for measuring an area's walkability. What we have done here towards that goal is to reveal how the distance traveled affects the impact of different kinds of establishments on the measured walkability. We also showed how the selection of different sets of amenities can alter scores and change which areas are measured as being more similar to others. Although there is no single, correct walkability score against which we can check our results, by revealing these sensitivities we uncover important factors to consider for any attempts at generating walkability scores and using them to compare areas by their walkability.

## References


1. Araki, S., Bramson, A.: Connecting the dots: integrating point location data into spatial network analyses. In: Benito, R.M., Cherifi, C., Cherifi, H., Moro, E., Rocha, L.M., Sales-Pardo, M. (eds.) COMPLEX NETWORKS 2020 2020. SCI, vol. 944, pp. 193–205. Springer, Cham (2021). https://doi.org/10.1007/978-3-030-65351-4_16
2. Bramson, A., Hori, M., Zha, B., Inamoto, H.: Scoring and classifying regions via multimodal transportation networks. Appl. Netw. Sci. **4**(1), 1–36 (2019). https://doi.org/10.1007/s41109-019-0191-7
3. Bramson, A., Okamoto, K., Hori, M.: Comparing methods for measuring walkability. Complex Systems Forthcoming (2021)
4. Calimente, J.: Rail integrated communities in Tokyo. J. Transp. Land Use **5**(1), 19–32 (2012)
5. Ellis, G., Hunter, R., Tully, M.A., Donnelly, M., Kelleher, L., Kee, F.: Connectivity and physical activity: using footpath networks to measure the walkability of built environments. Environ. Plann. B. Plann. Des. **43**(1), 130–151 (2016)
6. Hagberg, A., Swart, P., Chult, D.S.: Exploring network structure, dynamics, and function using NetworkX. Technical report, Los Alamos National Lab (LANL), Los Alamos, NM (United States) (2008)
7. keplergl: kepler.gl, August 2020. https://github.com/keplergl/kepler.gl
8. Koohsari, M.J., et al.: Validity of walk score® as a measure of neighborhood walkability in Japan. Prev. Med. Rep. **9**, 114–117 (2018)
9. Lüthy, M.: Japan-train-data, May 2017. https://github.com/adieuadieu/japan-train-data
10. Mapbox, February 2021. https://github.com/mapbox
11. OpenStreetMap contributors: Planet dump retrieved from https://planet.osm.org (2021). www.openstreetmap.org
12. Pedregosa, F., et al.: Scikit-learn: machine learning in Python. J. Mach. Learn. Res. **12**, 2825–2830 (2011)
13. Shimizu, C., Baba, H., Kawase, T., Matsunawa, N.: Walkability and real estate value: development of walkability index, June 2020. Online http://www.csis.u-tokyo.ac.jp/wp-content/uploads/2020/06/163.pdf

# Gaze Analysis on the Effect of Intervention on Ruminative Web Browsing

Giri Basanta Raj, Junya Morita(✉), and Thanakit Pitakchokchai

Shizuoka University, 3-5-1 Johoku, Naka-ku, Hamamatsu, Japan
giri.basanta.raj.19@shizuoka.ac.jp, j-morita@inf.shizuoka.ac.jp

**Abstract.** In the current highly developed information society, having a habit of rumination (repetitive and negative thinking) can be dangerous to our mental health. To prevent rumination during web browsing, the authors' previous study built an advertisement system that is regulated by a computational cognitive model and users' heart rate variability (HRV). To validate and extend the system, this paper presents analyses of behavioral and physiological data obtained from two studies where participants engaged in two successive tasks: mood induction and main tasks. Study 1 aimed to develop a detector for rumination distraction utilizing behavioral and physiological data. Owing to the SVM classification, a large contribution of gaze extracted from the facial movie was verified. To validate these findings, Study 2, in which a small number of participants engaged in the same tasks as Study 1, was conducted with a well-established eye-tracking system. Analyses of the gaze data obtained with this device confirmed high consistency with the data obtained from the facial movies, and also confirmed the influence of advertisements on participants' attention during web browsing. Summarizing the results of these studies, the current paper indicates the validity of distracting rumination by presenting prompts regulated with a personalized cognitive model.

**Keywords:** Rumination · Advertisement · Cognitive modeling

## 1 Introduction

Recently, we have been connected directly and indirectly with technology, which is now a part of our lives. Under such a situation, web browsing has become a daily life activity that positively and negatively influences our mood. In addition to the several advantages of this technology, we are prone to its negative effects as well. Among the several side effects of this technology, we focus on rumination, which is repetitive, negative thinking about one's unpleasant experiences caused by depressive moods. It has been revealed that high engagement with internet browsing is correlated with higher levels of anxiety [10]. Thus, with the


Y. Takama et al. (Eds.): JSAI 2021, AISC 1423, pp. 118–129, 2022.
https://doi.org/10.1007/978-3-030-96451-1_11

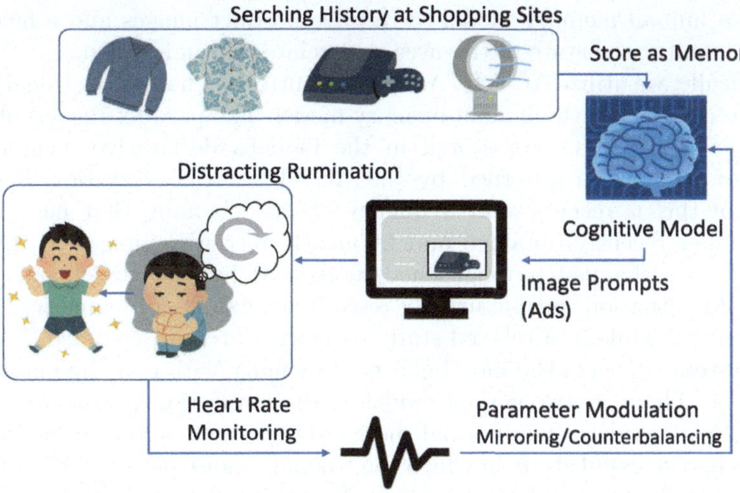


**Fig. 1.** System overview.

advancement of this technology, we suffer from mood disorders, such as depression, bipolar disorders, and anxiety, which have direct effects on our mood and emotions [12].

To address this negative side effects of technology, the discipline of affective computing, a multidisciplinary field spanning computer science, psychology, and cognitive science, is garnering attention to enable computations connected with human emotion [4]. Technologies in this field include recognizing the emotional state of a user with various types of sensing devices and response generation by performing specific, predefined product/service features [11]. Following such a method of affective computing, our previous study [8] developed a system to prevent rumination during web browsing. In this paper, we attempt to reveal the effect of the system on distracting rumination by analyzing gaze information.

The reminder of this paper comprises of a review of the system and descriptions of the two studies. The first study is a reanalysis of the data obtained in our previous study to construct a state detector related to the distraction of rumination. The second study is conducted to strengthen the findings obtained in the first study. In this study, we collect fine-grained gaze information in a small experiment. Summarizing these two studies, we discuss the effectiveness of our approach in the final section.

## 2 System of Model-Based Advertisement

Our previous study [8] developed a system of model-based advertisements that distracted ruminative web browsing. The basic concept is illustrated in Fig. 1. When the user ruminates, the system presents advertisements that divert attention from ruminating information on the web. To realize this concept, the system combines a human memory model to retrieve product images and a heart rate sensor to monitor the state of the user that relates to rumination.

Specifically, we utilize ACT-R (Adaptive Control of Thought-Rational) cognitive architecture [2] as the human memory model. The product images observed in the user's daily lives are stored in the model's declarative memory. The retrieval of memory is governed by the ACT-R's memory theory. The basic principle of this is recency and frequency effects, claiming that humans usually remember recently obtained and frequently retrieved images [1]; however, such effects are changed (strengthened/weakened) by the participants' mood. In a stressful situation, people usually retrieve a small set of memories, such as rumination [13] while in a relaxed state, memory search is expanded [5,7].

Our system connects the mood effects of memory with real-time monitoring of heart rate. There are two variant models of the connection: normal (NOR) and inverted (INV) models. The normal model exhibits ruminative behavior while the user is in a stressful state, in which the "memory noise parameter" (a random factor against the frequency and recency effects) is calculated corresponding to HRV. Thus, in the normal model, memory noise is assumed to be high when HRV is high, and vice versa. Inversely, the inverted model is designed to act as a counterbalance to the normal model that corrects rumination behavior. Therefore, the memory noise parameter of the inverted model is high when the HRV is low, and vice versa.

While users utilize the system, the ACT-R model retrieves the product images collected in their daily lives.[1] When the user enters a stressful state and ruminates (low HRV), the normal model tends to retrieve images that the user can recognize well (frequent recent images), while the inverted model tends to retrieve images that the user can barely remember. In contrast, when the user relaxes, the models behave inversely, the normal model tends to retrieve images that the user cannot recognize, and the inverted model retrieves images that the user knows very well. We assumed that the user ignores the images that he/she knows well and pays attention to/gets distracted by those that he/she has rarely or not seen before; we hypothesized that the inverted model will carry the relaxed state on as the user ignores the illustrated images and is distracted from rumination after observing the images.

The previous study [8] conducted an experiment in which two groups of participants utilized each system implementing the NOR/INV model and obtained supportive results indicating the effectiveness of the INV model in the subjective evaluation of distracting rumination; however, behavioral data clearly indicating this effect was not obtained. In this paper, we indicate further evidence of the effectiveness of the system by analyzing behavioral data, especially gaze information.

[1] We developed a browser extension to collect images on the browser when the user visits specific shopping sites (Amazon and Rakuten).

## 3 Study 1: Detector of Rumination Distraction

This study aims to construct a detector for the effect of the system introduced by the INV model. We consider that constructing such a detector is useful not solely to validate the effect of the system but also to develop a function that monitors users' rumination during web browsing. In the rest of this section, we will relay the results of this analysis after briefly describing the experiment conducted in the previous study.

### 3.1 Dataset

In the previous study [8], a behavioral experiment was conducted with 12 Japanese undergraduate students. The experiment comprised the following two sessions.

- Mood-induction task (MI): the participants recalled their recent bad memories and wrote the stories on a word processor software.
- Main task (MT): the participants searched for goods that can be useful for their future on the internet, while the system presented advertisements when the participants ruminated.

It was assumed that the first task induced a negative mood state (rumination), and that the second task caused the participants to recover from the negative mood state and induced a positive mood state (i.e., distracting rumination). The participants were seated in front of the computer during the experiment, and the face movement was recorded using a web camera. During the experiment, physiological arousal was also measured by a program that collected heart rate data from a heart rate monitor (Uniontool's myBeat). After the ACT-R model was activated in the MT session, the system displayed the image as an advertisement based on the user's current emotional condition.

### 3.2 Method of State Detection

**Measurement of Heart Rate and Eye Gaze.** In our study, the ACT-R models repetitively output an image every several seconds on the computer screen during the main task. We assumed that such a presentation of images implicitly intervenes in rumination caused by the mood-induction task. This study attempted to detect the effects of the intervention.

To detect this effect, the previous study [8] analyzed physiological arousal using heart rate, facial expression, and self-report questionnaires. Among them, eyes and their movements are assumed to be important in expressing a person's desires, needs, and emotional states [9]. The state and movement of eyes can also express people's cognitive and emotional states based on their content of interest [6]. Thus, this study focused on eye-gaze movement. Simultaneously, HRV was used as a signal to demonstrate the current emotional state.

Figure 2 illustrates an overview of the proposed method. The method was used to classify the normal and inverted models in both tasks using heart rate

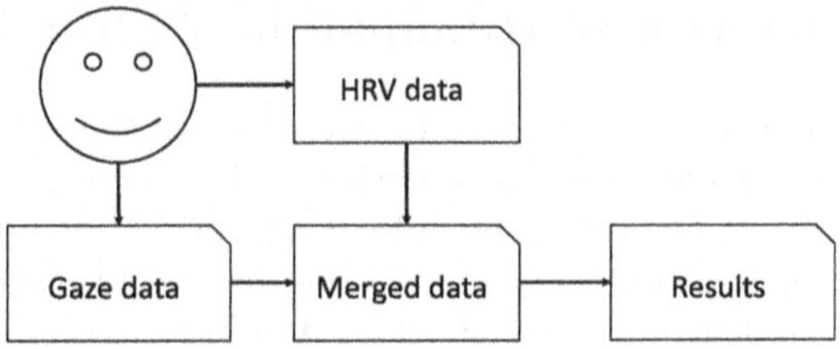


**Fig. 2.** Overview diagram of the state detection.

and eye gaze data. If the accuracy of classification between the two groups was higher in the main task than in the mood induction, we can conclude that the rumination of the participants was distracted by the advertisement on a computer screen in the main task.

**Data Preparation.** Among the two different data in Fig. 2, the HRV was measured utilizing Uniontool's MyBeat device and eye gaze data were extracted from the video with OpenFace software [3].

For each participant, 714 features were extracted from the recorded video using OpenFace. Out of them, we selected eight eye-gaze features. Generally, the participant's recorded video was 30 min long, of which approximately 15 min for the mood-induction task and approximately 15 min for the main task.

Thus, for each task and each participant, the 15-min long video contained 900 samples; each sample was 30 frames on average. The intervals between each heartbeat, that is, R-R interval data measured in the experiment, were also averaged for each second and concatenated with the gaze data. Each signal contained three beats. To resize the RRI data as the gaze data, we calculated the standard deviation of R-R intervals (SDNN) for each row in the data and merged this with the eye gaze data to make the final dataset and labeled the NOR and INV models.

**Classification Method.** We adopted support vector machines (SVMs) as a supervised machine learning algorithm. For both MI and MT, SVM classified the two groups where the participants were exposed to advertisements using the different models (the normal and inverted models). A leave-one-out five-fold cross-validation method was adopted with a linear kernel and 10 regularization parameters $C$ to calculate the accuracy of the classification between the results of the NOR and INV models in both tasks.

### 3.3 Results and Discussion

**Data Visualization.** Figure 3 illustrates examples of fluctuations of the HRV in the MI and MT. The red line in each graph indicates the HRV, while the dotted gray line represents the trend of HRV. From the figures, we can observe the

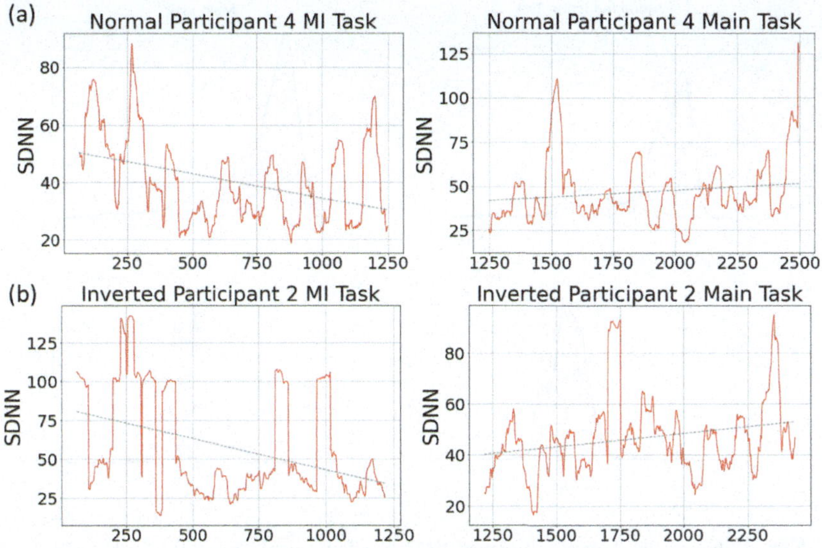


**Fig. 3.** Example of HRV data pattern in each group (a: Normal model b: Inverted model).

decreasing trends of HRV in the left panels (the mood-induction task) of Figs. 3 (a) and (b), indicating that the participants went into the negative thought (ruminated). However, in the MT, the participants' rumination gets distracted by advertisements more in the INV group than in the NOR group, as illustrated in Fig. 3 (b), although it highly depends upon the individuals.

Figure 4 illustrates the visualization of the data pattern of the MI and MT. Among the ten features of the dataset, SDNN and "gaze_0_x" were arbitrarily chosen to indicate the data distribution in each task. In Fig. 4 (a), while the NOR INV models have similar data patterns in the MI, they are distributed differently in the MT. Similarly, in Fig. 4 (b), the gaze movements for the NOR and INV models largely overlap in the MI, but we can observe the difference between the two peaks corresponding to the two models, suggesting the distraction of the rumination caused by the presentation of advertisements by the INV models.

**Classification Result.** Table 1 shows the accuracy of SVM classification of the two groups for each task. In this analysis, we applied two different feature sets, indicating the impact of eye movement and heart rate on the classification. Although all the accuracies in the table exceeded the chance level (.50) reflecting individual differences between participants in the NOR and INV groups, we can observe differences in accuracies between the two tasks for each feature dataset. This result indicates the possibility of detecting the effects of mood induction using our method.

We can also observe a larger difference between tasks in the gaze data only condition than the merged data, suggesting a large contribution of gaze pattern

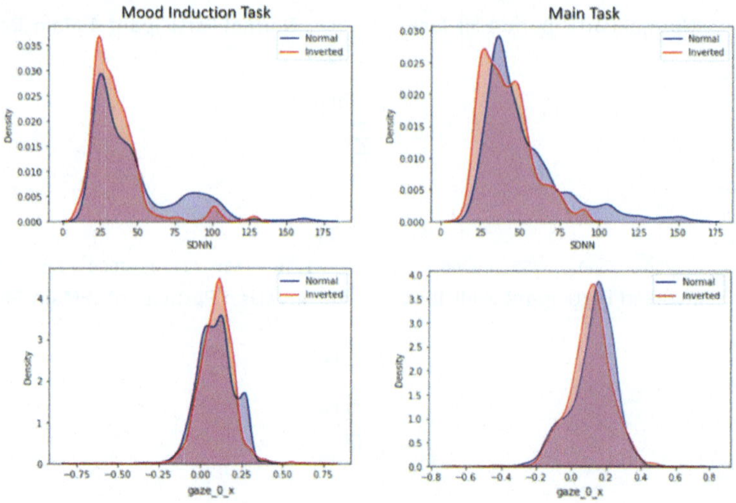


**Fig. 4.** Example density plots of SDNN (Top) and "gaze_0_x" (Bottom).

**Table 1.** Accuracy of mood-induction and main task

| Tasks | Data | Accuracy | Number of features |
|---|---|---|---|
| Mood-Induction task | Gaze data only | 0.66 | 9 |
| | Merged data | 0.72 | 10 |
| Main task | Gaze data only | 0.74 | 9 |
| | Merged data | 0.78 | 10 |

discriminating the two tasks. This result can be interpreted in an experimental setting. Because the mood-induction task made the participants write a text while recalling their recent unpleasant experiences, there seemed to be relatively fewer eye movements. However, in the main task, the participants utilized a web browser displaying advertisements on a computer screen. Thus, the eye gaze movements were assumed to be different, as they were prone to seeing the image displayed. Therefore, the classification accuracy of the gaze-data only condition is lower in the MI than in the MT.

In addition, we can observe that in the merged data (the eye gaze and the RRI data), the classification accuracy is higher than the eye gaze only data condition, indicating that the RRI also plays a role in the classification between the normal and inverted models. This result suggests some difference in recovering the mood induced in the mood-induction task; the inverted participants gradually became relaxed, while the normal groups tended to continue the mood induced in the previous task.

# 4 Study 2: Detailed Gaze Analysis

This study was conducted to support the results of the first study, following the experimental setup and experimental tasks with a few changes in the tasks. We utilized a remote eye-tracking device (Tobii pro nano) to monitor eye behavior in addition to a web camera. Using data obtained with this device, we conducted a detailed analysis of eye movements.

## 4.1 Data Collection

**Participants.** Four participants who are international master's degree students at Shizuoka University were asked to participate in the experiment. All of them reported that they usually visit shopping websites (Amazon) a few times a week.

**Design.** In the previous study, for the main task, the participants were divided into two groups that experienced different cognitive models: NOR and INV models. As indicated in Sect. 3.3, these two-model behaviors are distinguishable. Therefore, in this study, we conducted an experiment using solely the counter-balance model (INV model) to examine the actual effects of advertisements on the participants and gaze behavior during system use.

**Procedure.** As aforemantioned, the procedure of this study is approximately the same as the previous study, with a few changes in tasks and experimental devices. In this study, the mood induction task (MI) asked participants to recall unpleasant experiences and utilize the internet to search related thoughts instead of writing a report. This modification was made to mirror the previous mood induction task (MI), where the participants also engaged in an internet search for future life. We set up all data collection devices including Tobii's eye tracker. After confirming that all the devices worked well, the procedure was explained to the participants. The entire experiment took approximately 55 min.

## 4.2 Results

In this study, we present an analysis of gaze information to support the results obtained in the first study. Particularly, we indicate the difference in gaze patterns between the two tasks and the consistency of gaze information obtained with the Tobii eye tracker and OpenFace.

**Gaze Fixation on the Advertisement Area.** We obtained gaze point variables from the Tobii eye tracker. It extracted the gaze point variables as the raw gaze coordinates for each eye individually which are pixels display area coordinates system (DACS), which is a coordinate system with its origin in the top left corner of the screen (stimuli area), with Y pointing downwards. Accordingly, the gaze point variables extracted by the tracker are "Gaze point X [DACS px]," and "Gaze point Y [DACS px]" corresponding to the x and y-coordinates, respectively. The screen recording of the Tobii tracker presentation resolution was 1920 × 1080 (Pro Spectrum, Pro TX300), and the display resolution utilized for the experiment was 1920 × 1080. We calculated the total gaze point on the advertisement area for each task by calculating the x-coordinate between 1500 and 1800-pixels, and y-coordinate between 200 and 400 pixels which resembles the top right corner of the screen, that is, the advertisement area on a computer screen.

Table 2 presents the fixation on the screen and total fixation on the advertisement area for each participant. From this table, we can observe that the total fixation on the advertisement area in the main task is higher than that in the mood induction task.

**Table 2.** Total fixations and fixations on the advertisement area during the two tasks.

| | Mood induction task | | Main task | |
|---|---|---|---|---|
| | Total fixation | Ads area | Total Fixation | Ads area |
| P1 | 53840 | 96 | 53692 | 1782 |
| P2 | 54036 | 0 | 54032 | 674 |
| P3 | 54034 | 31 | 54028 | 82 |
| P4 | 53839 | 0 | 53942 | 1377 |

We also calculated the correlation between the total image displayed and the total fixation on the advertisement area in the main task. As explained in Sect. 2, the model-based advertisement system alters product images on the display based on the monitoring of users' HRV. The numbers of total and unique displayed images are indicated in Table 3, while the correlation between these numbers and the fixation on the advertisement area is indicated in Table 4. From the figure, we found a significant correlation between the total image displayed and total fixation in the main task. This result indicates that the participants' gaze movements were influenced by the advertisement.

**Correspondence Between Tobii and OpenFace.** As in the previous study, we utilized a web camera to record the video of the participants' face during the experiment and the OpenFace to extract the gaze data. For these video-based data, we examined the consistency with the data obtained using the eye tracker. The Tobii eye tracker extracted approximately 1000 samples per second, whereas OpenFace extracted approximately 30 samples per second. Therefore, we calculated the average of each second sample for each variable, outputting a total of 900 samples for a 15 min video.

Among the several variables obtained by both methods, we focused on the gaze direction. Tobbi eye tracker extracted the gaze direction variables as the unit vector for the direction of the gaze such as ["Gaze direction left X," "

**Table 3.** Number of total and unique displayed images.

| | Total images | Unique images |
|---|---|---|
| P1 | 158 | 70 |
| P2 | 127 | 91 |
| P3 | 85 | 59 |
| P4 | 140 | 88 |

**Table 4.** Spearman's correlation between images displayed and total fixation on advertisement area. Numbers in the parenthesis indicate p-value.

| | Total images | Unique images |
|---|---|---|
| MI | .32 (.68) | −.74 (.26) |
| MT | 1.00 (.00) | .20 (.80) |

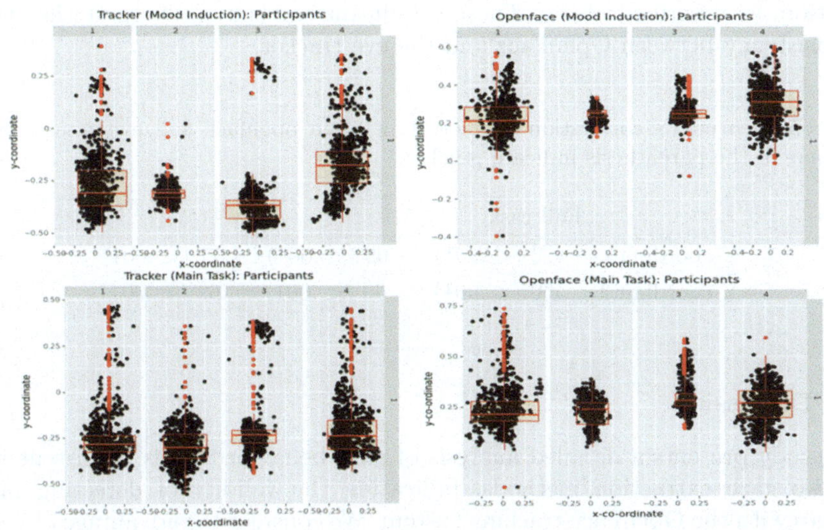


**Fig. 5.** Gaze direction of x-y coordinates in the two tasks for each participant (Top: mood induction task, bottom: main task; left: tracker, right: OpenFace).

Gaze direction left Y," "Gaze direction left Z," "Gaze direction right X," "Gaze direction right Y," "Gaze direction right Z"]. From this, for each participant in each task, we aggregated "Gaze direction left X" and "Gaze direction right X" to make x-coordinate and "Gaze direction left Y" and "Gaze direction right Y" to make the y-coordinate. Similarly, we aggregated the gaze direction of the OpenFace gaze direction to make the x and y-coordinates of the gaze direction.

We plotted the scatter and box for each task data of gaze direction in x-y coordinates, as illustrated in Fig. 5. From the figures, we observed several differences between the two gaze extraction methods. That is, the median of the y-coordinates of the tracker is negative, and that of OpenFace is positive. This is because of the different locations of each device (i.e., the web camera is at the top of the desktop, and the eye tracker is placed parallel to the bottom of the desktop). In addition, the tracker gaze direction moves toward the left, and the OpenFace gaze direction moves toward the right, reflecting different orientations of the OpenFace and eye tracker. Regarding the eye tracker, the orientation is the observer (i.e., participants who are sitting in front of a desktop), but in the case of OpenFace, the orientation is the web camera.

Despite these differences, the two extraction indicated similar individual differences. For example, both devices indicated that participant-2 (p2) and participant-3 (p3) have a negligible variance. To confirm the consistency between the two methods, we calculated the Pearson's correlation coefficient between the Tobii eye tracker and OpenFace gaze data. Table 5 presents the significant correlations between both x and y coordinates in all the tasks, although the correlation for the x-coordinate is negative because of the difference in the origin of the gaze direction, as described above. These significant correlations suggest that there is consistency between OpenFace and the eye tracker.

**Table 5.** Pearman's correlation coefficient between openface and eye tracker data. Numbers in the parenthesis indicate p-value.

| | Full task | MI | MT |
|---|---|---|---|
| X-coordinate | −.72 (< .001) | −.69 (< .001) | −.77 (< .001) |
| Y-coordinate | .60 (< .001) | .60 (< .001) | .62 (< .001) |

### 4.3 Discussion

This study presents a detailed analysis of gaze behavior that is consistent with the two gaze extraction methods, indicating the validity of detecting mood changes with the OpenFace tracking system. We consider the advantage of video-based gaze extraction (OpenFace) to include a system of model-based advertisements. The cost of introducing OpenFace is very low because users can utilize a face camera in most web browsing situations. The study presented here indicates the possibility of monitoring users' attention to ruminating/distracting information using this method.

## 5 Conclusion

The presented studies indicate the successful application of the proposed method of detecting rumination and distraction of rumination based on a computational cognitive model. The normal model was hypothesized to synchronize participants' rumination, while the inverted model was hypothesized to act as a counterpart to distracting participants' rumination by advertisements. The obtained results confirm these hypotheses. However, the data were limited to this study. In addition, further investigation of the application of this study and the limitations of the proposed method must be determined. In the future work, we will also develop a system integrating the advertisement system and the detector of rumination distraction developed in the first study.

## References


1. Anderson, J.R., Schooler, L.J.: Reflections of the environment in memory. Psychol. Sci. **2**(6), 396–408 (1991)
2. Anderson, J.R.: How Can the Human Mind Occur in the Physical Universe? Oxford University Press, Oxford (2007)
3. Baltrusaitis, T., Zadeh, A., Lim, Y.C., Morency, L.P.: OpenFace 2.0: facial behavior analysis toolkit. In: 2018 13th IEEE International Conference on Automatic Face and Gesture Recognition (FG 2018), pp. 59–66. IEEE (2018)
4. Calabrese, B., Cannataro, M.: Sentiment analysis and affective computing: methods and applications. In: Amunts, K., Grandinetti, L., Lippert, T., Petkov, N. (eds.) Brain-Inspired Computing, vol. 10087, pp. 169–178. Springer, Cham (2015). https://doi.org/10.1007/978-3-319-50862-7_13
5. Dancy, C.L., Ritter, F.E., Berry, K.A., Klein, L.C.: Using a cognitive architecture with a physiological substrate to represent effects of a psychological stressor on cognition. Comput. Math. Organ. Theor. **21**(1), 90–114 (2014). https://doi.org/10.1007/s10588-014-9178-1
6. He, H., et al.: Real-time eye-gaze based interaction for human intention prediction and emotion analysis. In: Proceedings of Computer Graphics International 2018, pp. 185–194 (2018)
7. Itabashi, K., Morita, J., Hirayama, T., Mase, K., Yamada, K.: Interactive model-based reminiscence using a cognitive model and physiological indices. In: Proceedings of the 19th International Conference on Cognitive Modeling, pp. 93–99. Pennsylvania State University, USA (2020)
8. Morita, J., Pitakchokchai, T., Raj, G.B., Yamamoto, Y., Yuhashi, H., Koguchi, T.: Regulating ruminative web browsing based on the counterbalance modeling approach. Front. Artif. Intell. (2022). https://doi.org/10.3389/frai.2022.741610
9. Ohno, T.: One-point calibration gaze tracking method. In: Proceedings of the 2006 Symposium on Eye Tracking Research and Applications, pp. 34–34 (2006)
10. Panova, T., Lleras, A.: Avoidance or boredom: negative mental health outcomes associated with use of information and communication technologies depend on users' motivations. Comput. Hum. Behav. **58**, 249–258 (2016)
11. Picard, R.W.: Affective computing: challenges. Int. J. Hum.-Comput. Stud. **59**(1–2), 55–64 (2003)
12. Valstar, M.: Automatic behaviour understanding in medicine. In: Proceedings of the 2014 Workshop on Roadmapping the Future of Multimodal Interaction Research including Business Opportunities and Challenges, pp. 57–60 (2014)
13. Van Vugt, M.K., Taatgen, N.A., Sackur, J., Bastian, M., Borst, J., Mehlhorn, K.: Modeling mind-wandering: a tool to better understand distraction. In: Proceedings of the 13th International Conference on Cognitive Modeling, pp. 252–257. University of Groningen Groningen, Netherlands (2015)

# Estimating Children's Intrinsic Motivation During a Class Based on Face Orientation and Gaze Information

Keisuke Morita[1], Natsuki Oka[1](✉), Kazuaki Tanaka[1], Masahiro Miyata[2], and Takashi Omori[2]

[1] Kyoto Institute of Technology, Matsugasaki, Sakyo-ku, Kyoto, Japan
nat@kit.ac.jp
[2] Tamagawa University, Tamagawagakuen, Machida, Tokyo, Japan

**Abstract.** Considering that non-cognitive abilities develop significantly during childhood, the Japanese early childhood education community has recently begun to focus on the study of non-cognitive abilities. However, measurement and estimation methods for non-cognitive abilities face quantitative and objective challenges. In addition, there is little research on the active use of artificial intelligence technology in education. Thus, this study proposed a method for quantitatively estimating concentration in the classroom, one of the children's non-cognitive abilities. Specifically, it examined the number of intersections between children's gazes using information on the direction of their faces and their gazes during class and determined whether all children should focus their gaze on a small area. Moreover, it quantitatively examined whether each child behaved appropriately in different situations by measuring the angle between each child's gaze and the center of gravity of the intersections of all the children's gazes. As a result, we found that we could capture the characteristics of each child's behavior.

**Keywords:** Non-cognitive abilities · Concentration · Attention · Behavioral data

## 1 Introduction

The development of non-cognitive ability is attracting attention worldwide because of its capability to survive in a rapidly changing society and enrich children's futures and lives. Non-cognitive abilities can be classified as follows [1]: (1) self-perceptions, (2) motivation, (3) perseverance, (4) self-control, (5) meta-cognitive strategies, (6) social competencies, (7) resilience and coping, and (8) creativity. It is believed that non-cognitive abilities grow significantly from infancy to lower grades of elementary school. However, there is insufficient knowledge about what non-cognitive abilities are and how they should be cultivated. In addition, intelligence quotient (IQ) and standard academic proficiency can be measured and quantified by tests, but methods for measuring and estimating non-cognitive abilities have problems in terms of quantity and objectivity.

Under these circumstances, Miyata, Morishita, et al. [2] and Miyata, Yamada, et al. [3] attempted to estimate the classroom's state (i.e., children's interest, concentration,


Y. Takama et al. (Eds.): JSAI 2021, AISC 1423, pp. 130–141, 2022.
https://doi.org/10.1007/978-3-030-96451-1_12

and divergence) by measuring children's gaze and face direction from camera images in the classroom. In this paper, we report the results of our attempt to judge whether a situation is one in which children should focus their gaze, and to estimate each child's willingness to participate in the class based on the information obtained from the gaze and facial orientation.

## 2 Behavioral Data of Children

The behavioral data used in this study were measured in an English class with 16 first-year students at Tamagawa Gakuen Elementary School on October 10, 2019. Two video cameras were installed at the top of the blackboard in front of the classroom. Each camera angle was set so as to capture all the seated children from the front. As these measurement data contain personal information, the measurement was conducted after clarifying the procedure to the guardians and obtaining their consent in advance.

We used NEC's remote gaze estimation technology [4] to detect faces and estimate gaze and face directions for each frame of the acquired video. The position of the target child's face was calculated by triangulation using the left and right cameras. For gaze and face directions, only the horizontal angle was used in this study. In this study, a total of 1 min and 9 s of the acquired video were used for the analysis.

The left half of Fig. 1 illustrates an example of children's facial orientations and their intersections. The white area in the figure represents the inside of the classroom, and the gray area on the upper right is the back of the classroom's front wall. The camera used for facial orientation and gaze recognition is located near the center of the front wall, and the purple square indicates its location. The green triangles indicate the children's positions with identification numbers for each child. The face directions are shown as straight lines. The intersections of the straight lines in front of the faces and within the classroom are shown as small red circles, and their center of gravity is shown as a big black circle. The right half of Fig. 1 shows children's gazes and their intersections at the same moment.

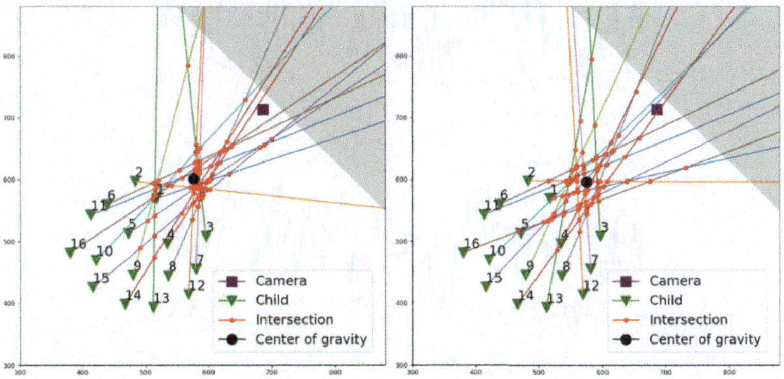


**Fig. 1.** (Left) Children's facial orientation and their intersections at the 293rd frame, and (Right) children's gaze directions and their intersections at the same moment

The aberrations in the optical system affected the triangulation and caused distortions in the positions of the children, as shown in Fig. 1. Specifically, the seats for children 1 and 2, 3–6, 7–11, and 12–16 were arranged in a horizontal row, whereas in Fig. 1, they were arranged in an arc. The distortion must have resulted in a backward shift in the center of gravity of the intersection points, as will be discussed later.

## 3 Analysis Methods

In this study, we quantitatively estimated the concentration in the classroom, which is one of the children's non-cognitive abilities. Specifically, we first examined the number of intersections between children's gazes using the information on the direction of their faces and their gazes during class and determined whether all children should focus their gaze on a small area. We then quantitatively examined whether each child behaved appropriately in different situations by measuring the angle between each child's gaze and the center of gravity of the intersections of all the children's gazes.

## 4 Results of the Analyses

### 4.1 Differences in Children's Level of Concentration Depending on the Scene

Figure 2 shows the shift in the children's concentration. The top half graph was created from the children's facial orientation data, and the bottom half was created from their eye-gaze data. The upper blue line shows the number of intersections of the children's face orientation, and the lower blue line shows that of their gaze. The upper red line indicates the average angle between the center of gravity of the intersections and the face orientation of each child. The lower red line represents the average angle of gaze.

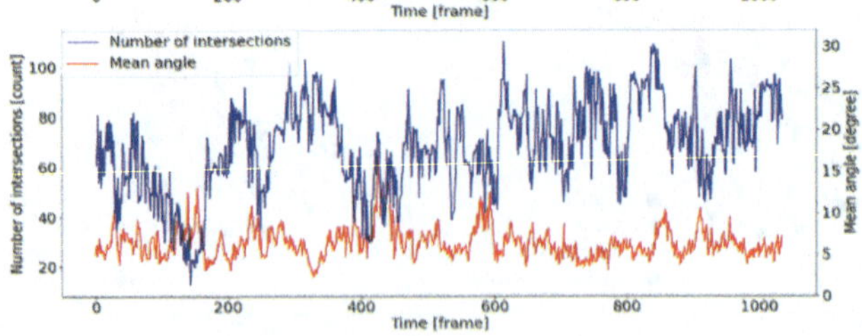


**Fig. 2.** The transition of children's concentration. The top half graph was created from the children's face orientation data, and the bottom half was created from their gaze data.

The upper and lower graphs show a similar trend in the number of intersections, as indicated by the blue line. A large number of intersections indicate that the gaze is focused on a small area and that this is when students should concentrate, while a small number of intersections indicate non-uniformity of gaze direction and the times when they can relax a little.

The red lines in Fig. 2 show that the gazes tend to be more concentrated near the center of gravity of the intersections than the face direction. This tendency could arise because their eyes may be looking in the proper direction, even if the face slightly deviates. Another possibility is that the gaze measurement may produce smaller angles than the face orientation measurement. For this reason, clarifying the cause of this tendency is a future task.

Figure 1 shows the gaze state at the 293rd frame where the number of intersections is large, whereas Fig. 3 shows the state at the 144th frame where the number of intersections is small. Finally, Fig. 4 shows the state at the 438th frame where the average angle of face orientation is quite large.

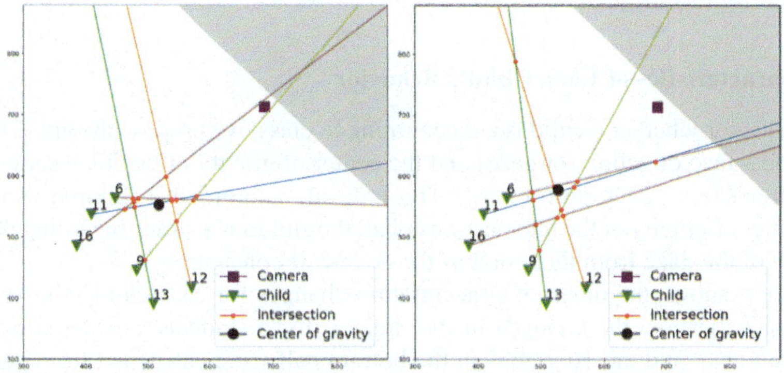


**Fig. 3.** (Left) Children's facial orientation and their intersections at the 144th frame, and (Right) children's gaze directions and their intersections at the same moment

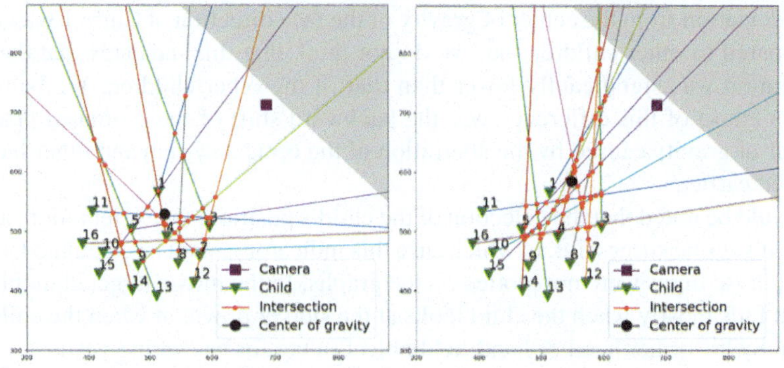


**Fig. 4.** (Left) Children's facial orientation and their intersections at the 438th frame, and (Right) children's gaze directions and their intersections at the same moment

The number of students displayed in Fig. 3 is small because objects in their hands hid their faces or they looked sideways or down; thus, their faces could not be recognized, and only the students who were recognized were displayed. As the number of students recognized is small, the number of intersections of face direction and gaze is also small.

Figure 4 displays data from a situation in which many of the children looked at the child beside them instead of at the front of the classroom. The left side of Fig. 4 shows that the center of gravity of the intersection points was behind Students 1 and 3. As a result, the angle between the face direction of Student 1 and the center of gravity became as large as 148°, even though he was facing forward. Consequently, the average angle of the 16 students exceeded 30°. The reason for this substantial angle is that the estimated position of the center of gravity of the intersections was shifted backward from its true position where the viewpoints were concentrated. As mentioned earlier, the leading cause of this misalignment seems to be the distortion of the students' estimated positions due to aberrations in the optical system. Another cause could be that due to angle or position measurement errors, the intersection points were shifted to the outside of the wall and were not counted as intersections, so the center of gravity was shifted backward from its original position.

## 4.2 Characteristics of Each Child's Behavior

As an indicator whether a child is concentrating in class, we propose the angle between each child's face direction (or gaze) and the center of gravity of the intersections of all the children's face directions (or gaze). This indicator assumes that children should gaze at the center of gravity of the intersections of all the children's gazes, regarding the slight deviation of the gaze from that point as the degree of concentration.

Figure 5 shows the index of concentration changes for each child. The 16 graphs in the top half show the changes in the indices of 16 students calculated from face orientation data, and the 16 graphs in the bottom half were calculated from gaze data. The identification numbers of the children correspond to those shown in Figs. 1, 3, and 4. The graphs show that the angle of face deviation tended to be greater than the angle of gaze deviation and that there was a difference in the magnitude of angles among children.

The deviation from the center of gravity of the face direction of Child 1 was substantial compared to other children, but we do not think that this indicates that the child's concentration was significantly lower than that of the other children. We believe that the major cause of this difference was the backward shift of the estimated position of the center of gravity caused by the aberration of the optical system and other factors, as mentioned earlier.

It should be noted that the detection of the child's position, face orientation, and gaze may not always be successful, in which case this indicator cannot be calculated; thus, the angles at those times were not plotted on the graphs. In the movie targeted in this paper, detection fails mostly when the child looks at the side or down, or when the child's face is hidden by his or her hand or hand-held object (a tag for answering yes/no).

In Fig. 5, the lack of a plot of the angle when the information was missing gives the impression that the deviation from the center of gravity was zero during that time. This impression is not suitable for understanding children's characteristics from charts.

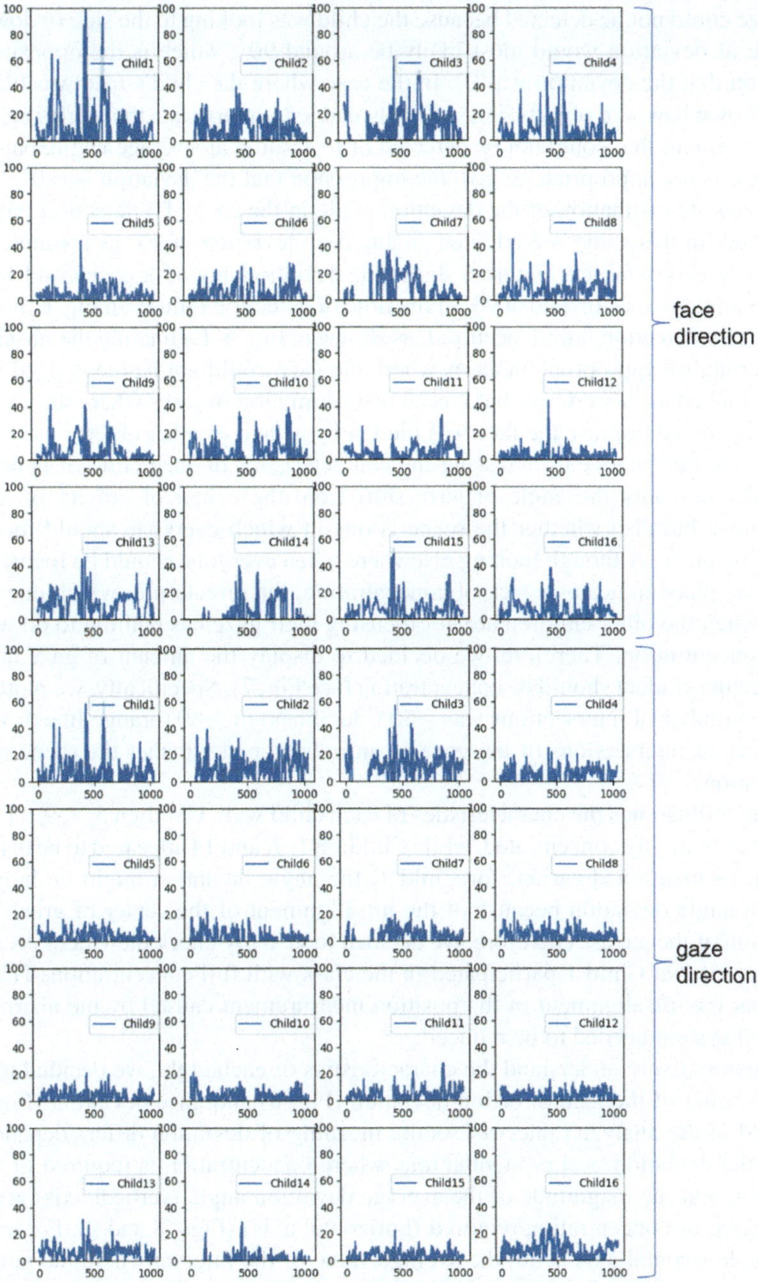


**Fig. 5.** The transition of the angle of the face direction (or gaze) of each child relative to the center of gravity of the intersections of the face direction (or gaze) of all children. The top 16 graphs were created from face orientation data, and the bottom 16 graphs were created from gaze data. Note that the angles when the face direction or gaze could not be recognized were *not* plotted.

If the gaze could not be detected because the child was looking to the side or down, then the angle of deviation would most likely be around 90°, which is the opposite of the impression that the deviation was 0°. In the case where the child's face was hidden by his or her own hands or objects, it seems reasonable to interpolate the angles before and after the segment that could not be detected or to assume an average angle, but even in this case, it is not appropriate to give the impression that the deviation was 0°.

An accurate estimation of the deviation angle in the case of a detection failure is a future issue. In this study, we adopted an angle of deviation of 90° as a simple way to deal with detection failure. Figure 6 shows the deviation angle for each child when the deviation angle was estimated at 90° at the time of detection failure. In Fig. 6, it is easier to see when a detection failure occurred, as shown in Fig. 5. Estimating the missing data as 90° is roughly appropriate in cases where the gaze could not be recognized because the child looked aside or down, but it is an overestimation in cases where the gaze could not be recognized because the face was hidden by a hand or other object.

From the perspective of assessing the characteristics of each child, it is necessary to consider not only the angle of gaze shift from the center of gravity of the gaze intersections, but also whether the scene is one in which everyone should focus their gaze on one place. Although looking elsewhere when everyone should be focusing their gaze in one place indicates a lack of concentration, the direction in which one child is looking when the other children are not focusing their gaze has nothing to do with that child's concentration. Therefore, we decided to display the amount of gaze deviation only when the student should be concentrating (see Fig. 7). Specifically, we plotted only when the number of intersections was ≥50 (blue lines) or ≥80 (orange lines). We used the number of intersections of the gaze as an indicator of whether the scene required concentration.

Figure 7 illustrates the characteristics of each child well. Children 5, 8, 9, 11, and 15 appeared to be highly concentrated, while Children 1, 7, and 14 appeared to be distracted. However, as mentioned earlier, for Child 1, the angle deviation might be larger than the actual angle deviation because of the misalignment of the center of gravity at the intersection of the gazes. Therefore, we decided to visually check the video. As a result, we confirmed that Child 1 participated in the class with full concentration. This result means that the misalignment of the position measurement caused by the aberration of the optical system needed to be reduced.

To quantitatively understand the characteristics of each child, we decided to obtain the mean value of the face direction deviation (Fig. 8) and gaze deviation (Fig. 9) for each child in the analyzed interval. As the meaning of deviation differs depending on whether the deviation occurs in situations where concentration is required or not, we set out to graph the magnitude of the average deviation angle (vertical axis) according to the degree of concentration required (horizontal axis) (Figs. 8 and 9). For example, where the horizontal axis is 50, the average value of the interval where the number of intersections is ≥50 is plotted. While Fig. 7 plots the deviation angle at each point where the horizontal axis is 50 and 80, Figs. 8 and 9 plot the average of these values over the time axis.

The curves in Figs. 8 and 9 are basically right-shouldered because as the threshold increases, the number of cases that are considered 90° due to lack of data decreases.

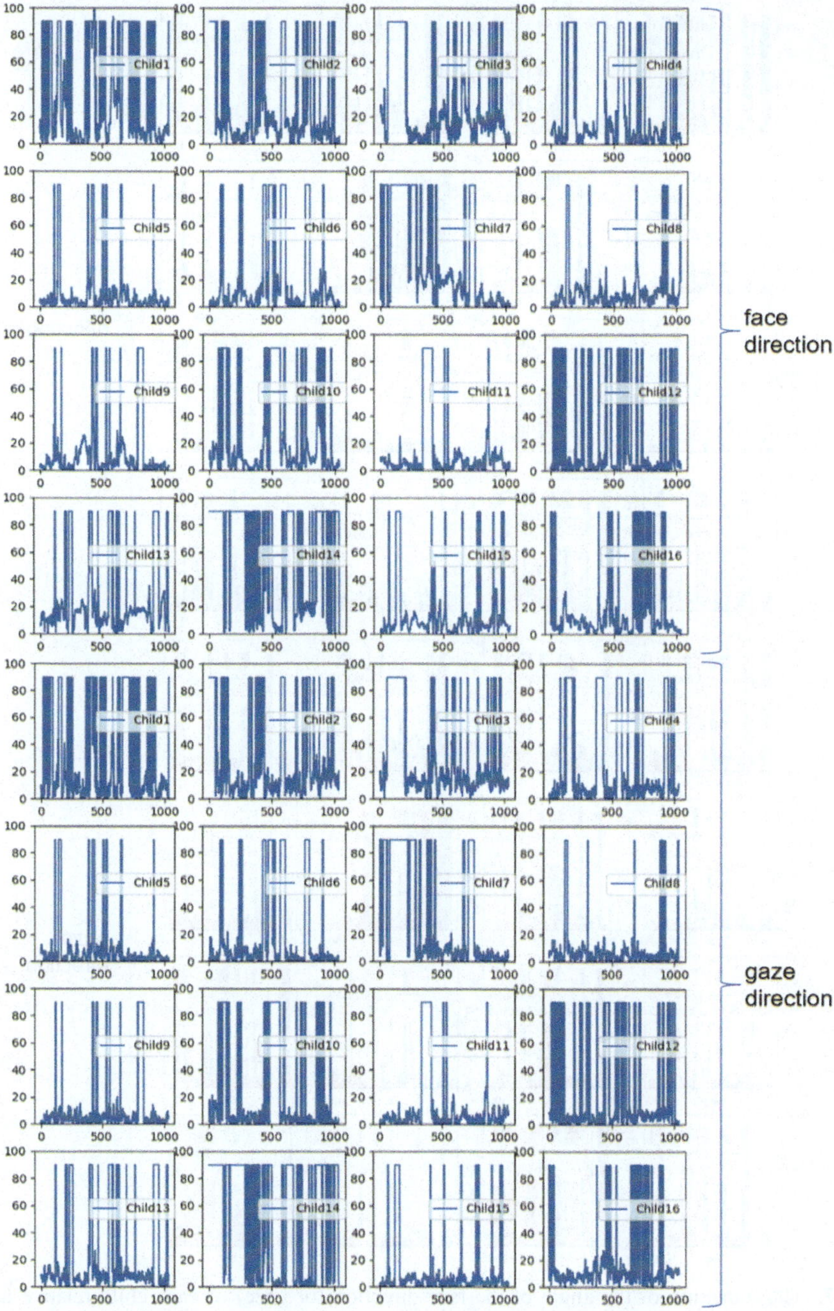


**Fig. 6.** The transition of the angle of the face direction (or gaze) of each child relative to the center of gravity of the intersections of the face direction (or gaze) of all children. The difference from Fig. 5 is as follows: The angles when the face direction or gaze could not be recognized were plotted as 90°.

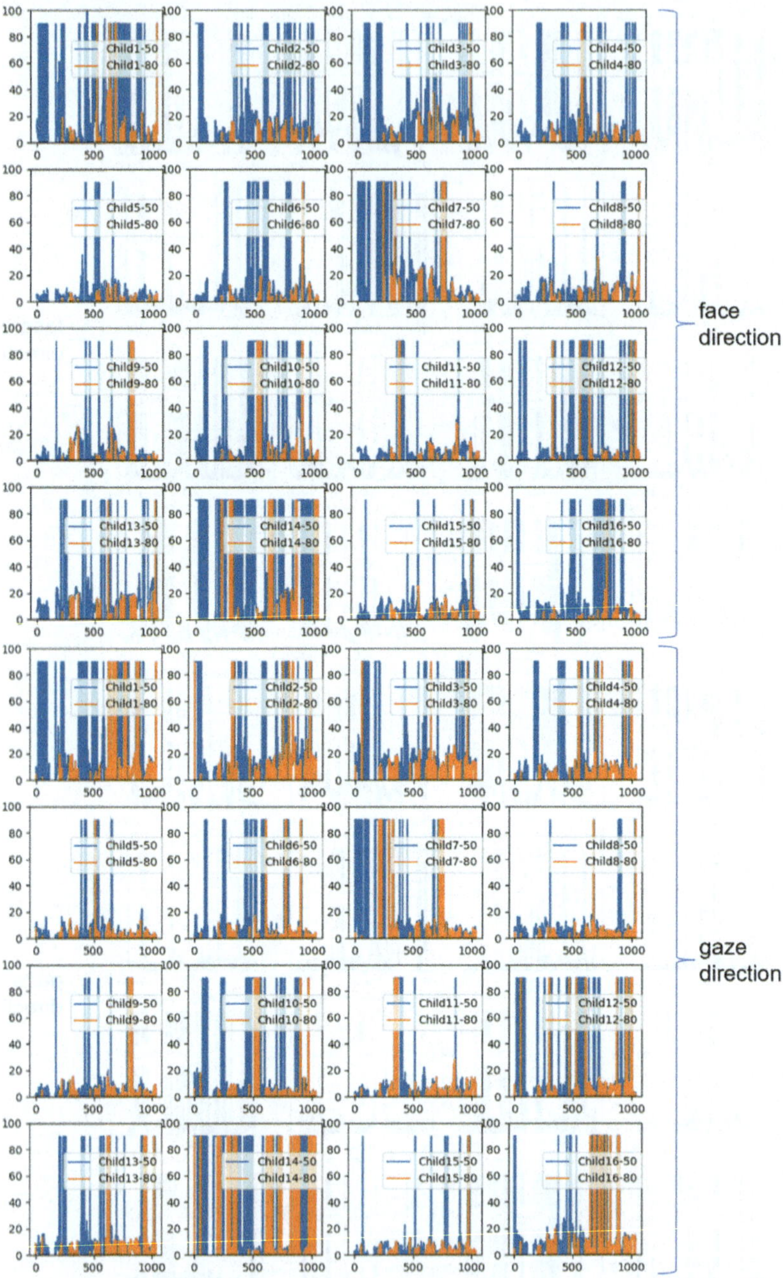


**Fig. 7.** The transition of the angle of the face direction (or gaze) of each child relative to the center of gravity of the intersections of the face direction (or gaze) of all children. The difference from Fig. 6 is as follows: Only the angles in situations where we presumed the children should concentrate were plotted. Precisely, we plotted only when the number of intersections was $\geq 50$ (blue lines) or $\geq 80$ (orange lines).

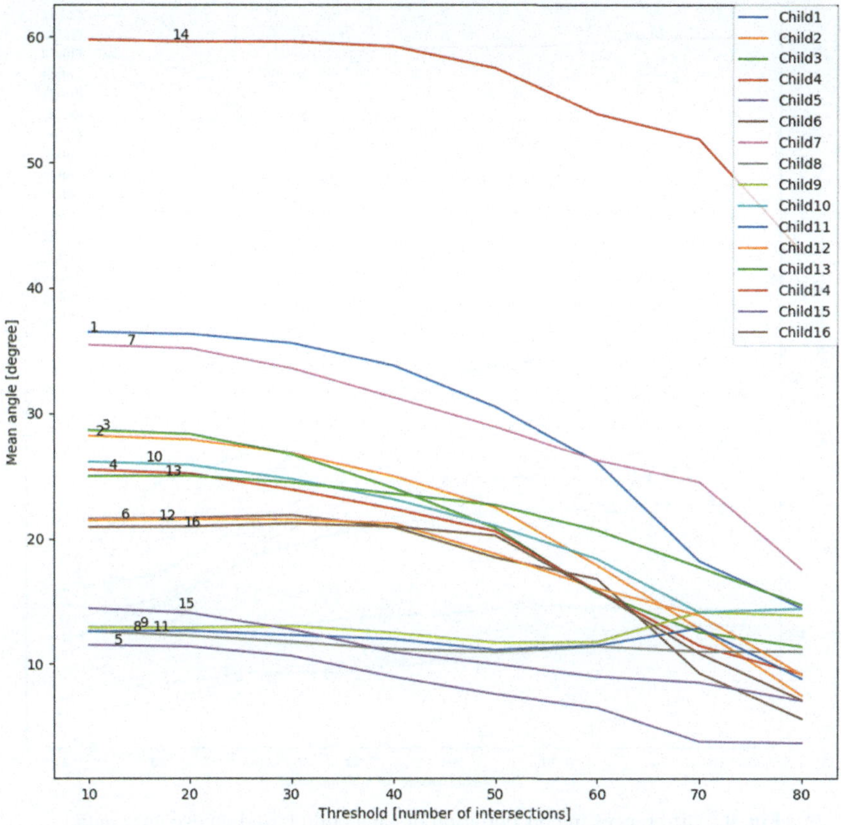


**Fig. 8.** Differences in concentration of each child based on face orientation data. The vertical axis is the average angle of deviation of the face direction of each child with respect to the center of gravity of the intersection points. The horizontal axis is the threshold for judging that the scene requires concentration.

The deviation angles of Child 14 were enormous; thus, we attempted to confirm this with the video data. We found that Child 14 was not often looking away, but he seemed restless and spent much time playing with the yes/no answer tag by holding it to the tip of his nose and spinning it around. For this reason, his face was often hidden by the tag and could not be detected, resulting in missing data, which was judged to be 90° during this period, and the average value of the deviation angle became large.

The reason for the larger deviation angle of Child 1 seems to be the distortion of the estimated position caused by the aberration of the optical system, as mentioned earlier. The video data did not indicate that the gaze of Child 1 tended to be off the point where it should be focused.

Children 5, 8, 9, 11, and 15, whose deviation angles were small, showed good posture and concentration in the class when checked on the video.

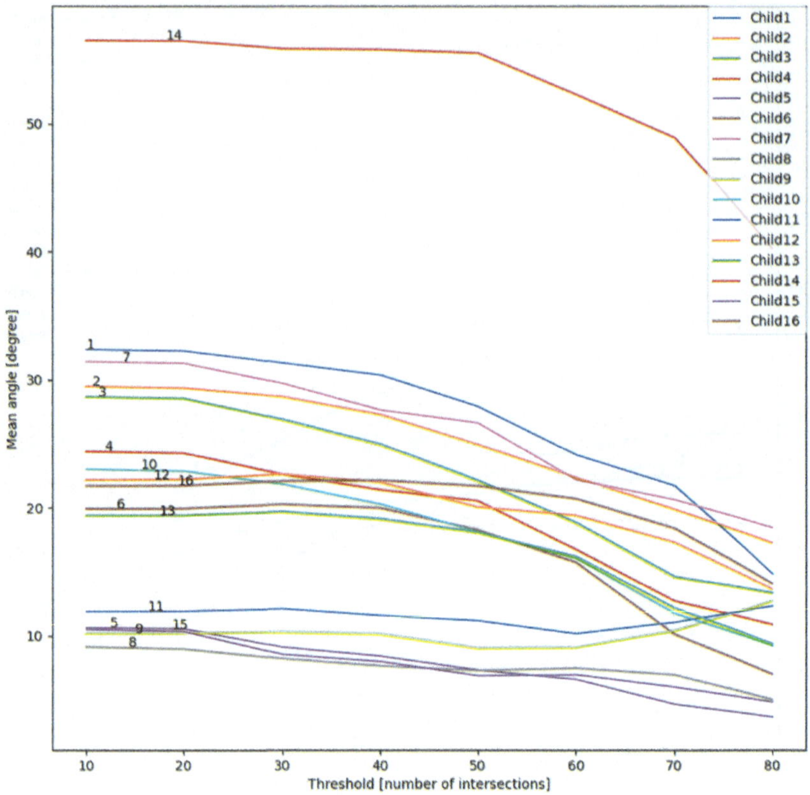


**Fig. 9.** Differences in concentration of each child based on eye gaze data.

## 5 Conclusion

This study proposed a method for quantitatively estimating concentration in the classroom, one of the children's non-cognitive abilities. Specifically, the number of intersections between children's gazes during class were examined, and we estimated whether all children should focus their gaze on a small area. We also quantitatively estimated each child's willingness to participate in class by measuring the angle between each child's gaze and the center of gravity of the intersections of all the children's gazes. As a result, in many cases, we could capture the characteristics of each child's ability to concentrate.

Appropriate handling of missing data is an essential task for the future. Missing data are caused by a failure to detect a child's face, which can occur in at least two cases: when the child looks sideways or down, or when the child's face is hidden by a hand or other object. For an accurate understanding of children's characteristics, these cases need to be treated differently.

Another future issue is the distortion of the position estimation due to aberrations in the optical system. We consider that the distortions caused a shift in the face orientation and gaze intersections, which affected the accuracy of the estimations of children's concentration. Although we did not take actual measurements of the desk positions at the time of data acquisition and cannot make a quantitative evaluation, we believe that the effect on Child 1 was significant, and the effects on the other children were not. To address this issue, we have started to take new data by replacing the lens with one with less distortion, which will improve the accuracy of position estimations.

**Acknowledgment.** This work was supported by JSPS KAKENHI Grant Number JP19H01718 and by the Canon Foundation Research Grant Program "Science and Technology that Achieve a Good Future".

## References

1. Gutman, L., Schoon, I.: The impact of non-cognitive skills on outcomes for young people: literature review. Education Endowment Foundation, London (2013)
2. Miyata, M., Morishita, Y., Yamada, T., Takamoto, M., Imaoka, H., Omori, T.: Estimation of child's interest target using face information: trial of field sensing toward cognitive science in the classroom. In: The 37th Annual Meeting of Japanese Cognitive Science Society (2020). (in Japanese)
3. Miyata, M., Yamada, T., Takamoto, M., Morishita, Y., Omori, T.: Estimation of student's characteristics during classroom activities using face information. In: The 38th JSET Spring Conference (2021). (in Japanese)
4. Takamoto, M., Morishita, Y., Imaoka, H.: An efficient method of training small models for regression problems with knowledge distillation. arXiv:2002.12597v1 (2020)

# Recognising Flexible Intents and Multiple Domains in Extended Human-Robot Dialogues

Graham Wilcock(✉)

CDM Interact, Helsinki, Finland
graham.wilcock@cdminteract.com
https://www.cdminteract.com

**Abstract.** We focus on extended multi-turn multi-topic human-robot dialogues, which can be more challenging than short question-answering interactions. In earlier work we developed two dialogue systems for Nao robots: WikiTalk supporting Wikipedia-based dialogues on open-domain topics, and CityTalk supporting task-based dialogues on restaurant and hotel domains. We used WikiTalk with ERICA at Kyoto University, and now wish to make our systems available on multiple robot platforms. To support this aim we use Rasa open-source conversational AI, which creates transformer-based dialogue models that aim to recognise flexible intents in multi-turn multi-domain dialogues. To improve CityTalk we use Rasa knowledgebase actions backed by Neo4j graph databases which support knowledge graphs for multiple domains. By adding taxonomies and other semantic context to the knowledge graphs we aim to give more intelligent dialogue responses. We also plan to use the large MultiWOZ multi-domain dialogue dataset to support additional domains.

**Keywords:** Extended human-robot dialogues · Multi-domain dialogues · Transformer-based dialogue models · Flexible intent recognition

## 1 Introduction

Our work focusses on extended multi-turn multi-topic human-robot interaction. Extended dialogues that continue for 10 min or more pose different challenges from short question-answering interactions, as dialogue history and the ongoing changing context will influence subsequent responses.

In earlier work we developed two dialogue systems for Nao robots. WikiTalk [6,15] supports open domain multi-turn multi-topic dialogues using Wikipedia, changing smoothly between many topics during the dialogue. CityTalk [16,17] supports task-based multi-turn dialogues about restaurants and hotels, changing smoothly between these two domains during the dialogue.

We now wish to make these systems available not only on Nao robots but across multiple robot platforms. We also aim to improve CityTalk with better


Y. Takama et al. (Eds.): JSAI 2021, AISC 1423, pp. 142–153, 2022.
https://doi.org/10.1007/978-3-030-96451-1_13

semantic context using knowledge graphs for the restaurant and hotel domains, and to extend CityTalk beyond restaurants and hotels to multiple domains.

To support these aims we use Rasa open-source conversational AI [2]. Rasa supports transformer-based dialogue models [14] which aim to recognise flexible intents in extended multi-turn dialogues across multiple domains, not merely recognising pre-specified intents within a single domain. This is attractive for our aims. We combine Rasa with a Neo4j graph database [10] in order to build knowledge graphs enriched with semantic context including taxonomies [1]. We also hope to use the MultiWOZ multi-domain dialogue dataset [3] in order to extend CityTalk to more domains.

The paper is structured as follows. Section 2 summarizes our previous work on WikiTalk and CityTalk. Section 3 presents the Rasa conversational AI framework and its use of transformer-based dialogue models. Section 4 describes the use of knowledgebases with Rasa, including knowledge graphs supported by Neo4j graph databases. Section 5 introduces the MultiWOZ multi-domain dialogue corpus. Section 6 concludes and describes future work.

## 2 WikiTalk and CityTalk

In previous work we developed WikiTalk and CityTalk for Nao robots. We briefly summarize that work here and give links to videos and further details.

### 2.1 WikiTalk: A Talking Wikipedia

With WikiTalk [6,15] robots can talk fluently about thousands of topics using information from Wikipedia, shifting smoothly from topic to topic following the user's interests when the user mentions the names of related topics.

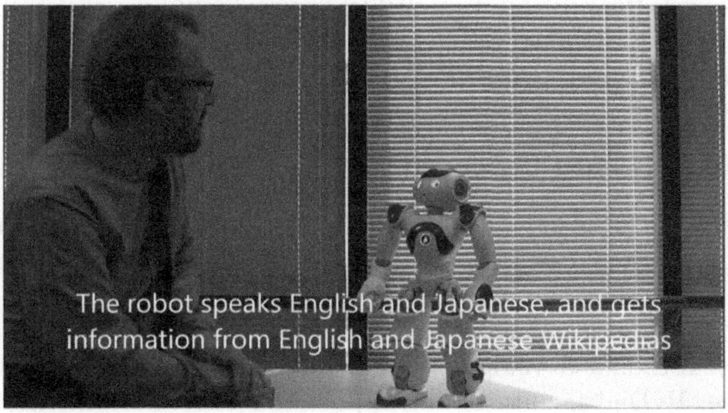


**Fig. 1.** Multilingual WikiTalk at Doshisha University.

WikiTalk is multilingual, currently with localizations for English, Finnish and Japanese. Figure 1 shows a screenshot from a video[1] in which a Nao robot talks

[1] Multilingual WikiTalk: https://www.youtube.com/watch?v=NkMkImATfYQ.

about Shakespeare and other topics for about 15 min in English and Japanese with an English user and a Japanese user, switching languages when the users say *English* or *Nihongo*.

WikiTalk uses the changing dialogue context to disambiguate ambiguous topic names. For example, when one of the users says *Julius Caesar* while the robot is talking about Shakespeare, it is understood to refer to the play by Shakespeare, but when the robot is subsequently talking about the play, *Julius Caesar* is taken to refer to the character in the play [18].

**Fig. 2.** ERICA and WikiTalk at Kyoto University.

We also used WikiTalk with the lifelike female android robot ERICA at Kyoto University [7]. Figure 2 shows a screenshot from a video[2] in which ERICA talks for around 10 min about multiple topics including classical languages, autonomous robots, artificial intelligence, the concept of intelligence in medieval philosophy, android robots and gynoid robots.

### 2.2 CityTalk: Local Information for Tourists

With CityTalk [16,17] Nao robots tell tourists about local hotels and restaurants. The robot can switch between the two domains (hotels and restaurants) on request during the dialogue using a topic tracker.

CityTalk was the first robot dialogue system to use PyDial [11], a statistical dialogue system toolkit developed by the Dialogue Systems Group at University of Cambridge. PyDial facilitates the application of reinforcement learning and other deep learning techniques in dialogue modelling.

Domain-specific knowledge about hotels, restaurants and other application domains is stored in SQL databases. In a short video[3] of the first demo of CityTalk the robot talks about hotels and restaurants in Cambridge using the

[2] ERICA and WikiTalk: https://www.youtube.com/watch?v=Aq4Rfwrktr0.
[3] CityTalk Cambridge: https://www.youtube.com/watch?v=zWdd7kv5sX8.

existing databases from the PyDial toolkit. We created new databases for hotels and restaurants in Tokyo.

**Fig. 3.** CityTalk: Tokyo 2020 hotels and restaurants.

Figure 3 shows a screenshot from a video[4] of an extended multi-turn dialogue in which a Nao robot talks about hotels and restaurants in the Tokyo Waterfront area near the Tokyo 2020 Olympics. First the robot gives information about three hotels in different price ranges. When the user asks about restaurants, the robot switches domains from hotels to restaurants and talks about several restaurants in the same area in different price ranges serving several different types of food. The extended dialogue lasts about 10 min.

The architecture of PyDial is shown in Fig. 4. The dialogue agent contains a pipeline of 4 components: Semantic Parser, Belief Tracker, Dialogue Policy (System Reply), and Semantic Output. In addition, a Topic Tracker detects when the user changes from one domain to another.

PyDial is a text-based system. For spoken dialogues, an external speech client can provide speech recognition (ASR) and speech synthesis (TTS) services as shown in Fig. 4. CityTalk uses the speech components of the Nao robot.

## 3 Rasa Conversational AI Framework

To support our aim to make WikiTalk and CityTalk available on multiple robot platforms, we now use Rasa open-source conversational AI framework [2]. Like PyDial, Rasa is committed to exploiting deep learning for dialogue modelling, but Rasa makes it easier for non-specialist software developers to build advanced dialogue systems by including the newest results from deep learning research within standard Rasa components.

Most spoken dialogue systems are currently based on underlying text-based dialogue systems, to which they add speech recognition and speech synthesis

[4] CityTalk Tokyo 2020: https://www.youtube.com/watch?v=OhjIJp8XBEA.

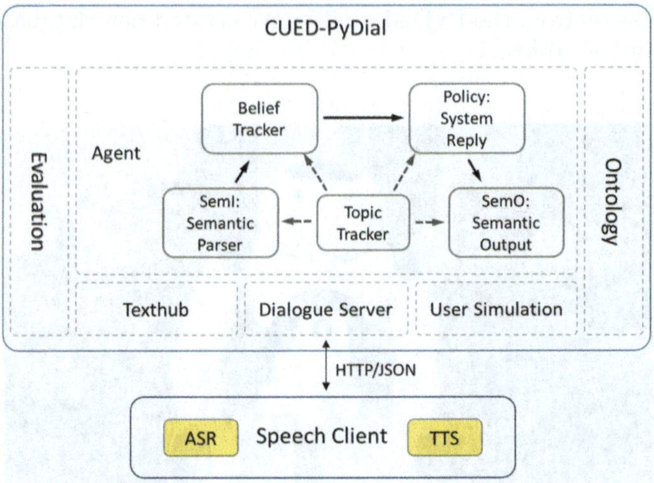


**Fig. 4.** PyDial architecture overview, from [11].

interfaces. Like PyDial, Rasa is a text-based system. To use Rasa with speech, external speech recognition and speech synthesis components must be provided. In the architecture overview[5] in Fig. 5, speech processing must be done by the Input/Output Channels between the user and the Rasa Agent.

Like the PyDial architecture in Fig. 4, most text-based dialogue systems have a pipeline of four components which perform natural language understanding (NLU), dialogue state tracking, dialogue policy (to select the system response), and natural language generation (NLG). To exploit advances in deep learning, some neural dialogue models merge all the non-speech components (NLU, state tracking, dialogue policy and NLG) into a combined end-to-end dialogue model. While Rasa certainly exploits deep learning, it keeps the NLU component (Rasa NLU) separate from the other dialogue management components (Rasa Core).

In Fig. 5, Rasa NLU is labelled NLU Pipeline and Rasa Core is labelled Dialogue Policies. One advantage of this separation for multilingual systems is that Rasa NLU can focus on language-specific requirements, recognizing Intents and Entities in user utterances in each language, while Rasa Core can focus on other dialogue management requirements using language-neutral representations of Intents and Entities.

Access to external services such as databases is supported via custom actions which can be written in Python. Custom actions are executed by Action Server, part of Rasa SDK shown in Fig. 5.

### 3.1 Transformer-Based Natural Language Understanding

In Rasa, natural language understanding is performed by the NLU Pipeline component shown in Fig. 5. The pipeline can be configured to include whatever

[5] Rasa architecture figure: https://rasa.com/docs/rasa/arch-overview.

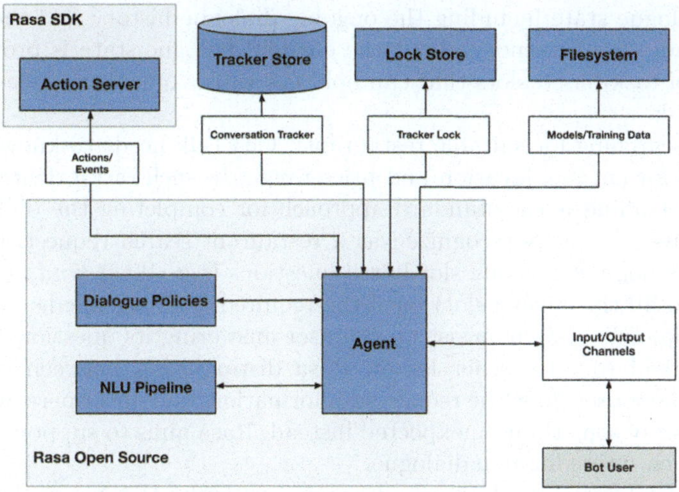


**Fig. 5.** Rasa architecture overview (from Rasa Docs)

sequence of NLU components is best for the system task. The two main tasks that the NLU pipeline must perform in a dialogue system are intent classification and entity recognition. These two tasks can be performed by separate components, but Rasa has developed an original transformer-based component that performs both tasks in an integrated process. This is the Dual Intent and Entity Transformer (DIET) [4].

Recently large pre-trained language models like BERT have become widely used in NLU tasks, but such general-purpose models trained mainly on written texts may not fit a conversational AI application for a specific specialised domain, especially for a dialogue system using spoken language.

Rasa strongly encourages Conversation-Driven Development, in which the system is continuously improved from the start using real-world conversations with end users. This requires that the language models can be re-trained quickly. Fine-tuning very large pre-trained models with your own conversational data is possible, but may be too slow for an iterative development workflow.

The problem with getting started is that there is a lack of conversational data at first. The DIET transformer can incorporate pre-trained word embeddings to help where user data is sparse, but it is claimed [4] to achieve a high performance using only sparse word and character level n-gram features, even without pre-trained embeddings, while being much faster to train than fine-tuning BERT.

Another advantage of being able to work with sparse data is that large pre-trained models are not currently available for most of the world's languages.

### 3.2 Transformer-Based Dialogue Policy

In Rasa, the task of deciding the system response is performed by the Dialogue Policies component shown in Fig. 5. The appropriate response depends on the current dialogue state including the ongoing dialogue history and the changing context. Short-term memory about the current dialogue state is provided by a collection of task-based slots that can hold the values of features relevant to the dialogue task.

Before searching for suitable restaurants, CityTalk needs to know the user's preferences for cuisine, location and price range. In such task-oriented dialogue systems, slot-filling is the standard approach for completing the dialogue task. When the user intent is recognized as a restaurant search request, the system drives the dialogue by asking slot-based questions (e.g. *What kind of food would you like?* to fill the cuisine slot) until the required slots are filled.

Some dialogue systems insist on the user answering its questions before the user is allowed to raise other issues. Rasa distinguishes between cooperative dialogue, where users give the requested information, and uncooperative dialogue where users say something unexpected instead. Rasa aims to support systems in handling such unpredictable dialogue.

The PyDial team used recurrent neural networks (RNNs) to build multi-domain dialogue state trackers, and showed that models trained on combined domains can outperform models trained on a single domain [9]. However, RNNs assume that every item in a sequence is relevant for encoding the sequence, which may be true in cooperative task-oriented dialogue but not in chitchat or uncooperative dialogue where meta-level utterances (e.g. *Why do you ask that?*) may be interleaved with slot-filling.

Rasa developed a Recurrent Embedding Dialogue Policy (REDP) [13] to overcome this limitation of RNNs. The aim is to build multi-domain dialogue state trackers that can handle the interleaving of task-related and non-task conversation, and recover gracefully from unprompted chitchat from users. A key idea is to generalize across domains by transferring learned behaviours from one domain's slot-filling task to other domains.

However, the success of transformer-based attention models [12] has led Rasa to replace the REDP policy by a newer Transformer Embedding Dialogue (TED) policy [14]. The TED policy uses attention to select which items in a dialogue sequence are relevant for the given task, and which items can be ignored as irrelevant (or relevant for non-task goals such as chitchat).

By using the TED policy Rasa should be able to recognise flexible intents in extended multi-turn dialogues that switch between multiple domains. This will be a significant advance on recognising pre-specified intents within a single domain, and makes Rasa attractive for our aims.

## 4 Knowledgebases and Knowledge Graphs

In addition to enabling developers to write standard custom actions in Python for access to external services such as SQL databases, Rasa supports a special type of action for access to knowledgebases. These knowledgebase actions[6] are currently an experimental feature.

[6] Rasa knowledgebase actions: https://rasa.com/docs/action-server/knowledge-bases.

Rasa knowledgebase actions can be used immediately with small in-memory knowledgebases in JSON format that can be loaded directly from a JSON file. With additional programming, they can also be used with large knowledgebases and more complex knowledge graphs stored in external graph databases. These two cases are described in the next two sections.

### 4.1 Rasa Knowledgebase Actions

At first there were some problems with the experimental knowledgebase actions. In order to go ahead quickly using Rasa for CityTalk, we wrote standard custom actions to search the existing SQL restaurant and hotel databases that we had created for use with PyDial. When the initial problems were fixed, we extracted a subset of restaurants and hotels from our SQL databases to CSV and created a small knowledgebase in JSON format. We switched from domain-specific custom actions to knowledgebase actions in order to develop a more domain-independent approach to accessing world knowledge.

With the standard custom actions we had used a separate custom action and a separate SQL database for each domain (a hotel custom action to query a hotel database and a restaurant custom action to query a restaurant database). With the knowledgebase actions approach, we use a single knowledgebase in which restaurants and hotels are objects of two different types, and we use a single knowledgebase action that can make queries about objects of any type.

The current experimental knowledgebase actions in Rasa support two basic query types. The first query type searches the knowledgebase for objects of a given type that match a given set of properties (for example, find restaurants with a moderate price range and serving Japanese cuisine). The second query type gets a requested property of an already-mentioned object (for example, get the phone number of the mentioned restaurant). These two generalized query types between them provide support for most of the wide range of user queries asked in practical information-seeking dialogues.

This relatively elegant approach, based on a single knowledgebase including objects of different types and a single knowledgebase action that queries objects of all types, offers potential advantages for multi-domain dialogue systems. To add new domains does not require creating new databases and new custom actions, it requires only adding new object types to the existing knowledgebase. The single knowledgebase action will then make its two types of queries about the new object types. Of course, with either knowledgebase actions or standard custom actions, the NLU component needs to be retrained to recognise entities of the new types in user utterances about the new domains.

A further potential advantage of using a single knowledgebase query action is that it may provide a more consistent interaction experience for the user when switching between different domains, compared with switching between different custom query actions that may each have a different interaction style.

### 4.2 Neo4j Graph Databases and Knowledge Graphs

After developing Rasa knowledgebase actions for CityTalk with a small-scale in-memory knowledgebase in JSON format, we imported the data into a Neo4j graph database [10]. Like other database systems, data can easily be loaded into Neo4j from CSV files. We deploy the database both locally with Neo4j Desktop and in the cloud with a Neo4j AuraDB database server endpoint.

Moving to a graph database like Neo4j requires that the Rasa knowledgebase actions must be provided with a new interface module that can perform queries correctly in the graph database. The example knowledgebase actions currently provided by Rasa only search in-memory knowledgebases in JSON format. We made a prototype interface module that can query our Neo4j graph database. The most important requirement is that the queries must follow the relations between nodes in the graph. For example, the cuisine served by a restaurant is not simply a property attribute of the restaurant node. Different cuisines are themselves nodes in the graph, and restaurant nodes are connected to cuisine nodes by a specific relation, in this case the *serves* relation.

The motivation for moving the knowledgebase to a graph database is not only to allow scaling up in quantity to much larger numbers of restaurants and hotels, but more importantly to allow upgrading the quality of the information. Neo4j graph databases can efficiently support semantically-enriched knowledge graphs [1], and we intend to add enhanced semantic contexts in order to provide better dialogue responses and more intelligent interaction with the user.

For example, in the PyDial version of CityTalk the cuisine that a restaurant serves is simply a string-valued property of the restaurant. If the database has three restaurants in a requested location with cuisines specified as *French*, *Italian* and *European* respectively, and if the user asks for *European food*, only the third restaurant is found as a match. Such a system does not know that *French* and *Italian* cuisines are subtypes of *European* cuisine, and the first two restaurants are ignored. Worse, if there is no restaurant with cuisine specified as *European*, the user is incorrectly told that there is no restaurant serving European food in the requested location. With a knowledge graph this problem can be solved by adding a taxonomy of *subtype-of* relations and by providing queries that can follow those relations.

We plan to enrich the knowledge graph incrementally by adding semantics in the form of hierarchies of different kinds. For example the desired location of a hotel or restaurant can be specified by the user in many different ways including an informal district name (*in Odaiba*), an official city or ward name (*in Koto City*), or more vaguely (*anywhere in Tokyo*). These place names may be recognized by the NLU pipeline as location names, but the system will treat them as different locations if they are merely strings. The system can give more intelligent responses if it has a hierarchy of locations in which (most of) Odaiba is in Koto City, and Koto City is part of Tokyo. This kind of hierarchy can be added to the knowledge graph in the Neo4j database by adding a layer of *part-of* relations.

## 5 MultiWOZ Dialogue Corpus

To pursue our aim to extend CityTalk to more than two dialogue domains we intend to use MultiWOZ [3,5], a freely available multi-domain Wizard-of-Oz dataset for task-oriented dialogue modelling[7].

MultiWOZ is multi-domain, with dialogues on hotels, restaurants, taxis, trains, local attractions, hospitals, and police stations, a total of 7 domains. Many of the dialogues include more than one domain, and in some dialogues users change their goal in the middle of the conversation. The conversations in the corpus are multi-turn dialogues, averaging about 14 turns per dialogue. This matches our focus on extended dialogues.

MultiWOZ contains over 10,000 task-oriented dialogues and is an order of magnitude larger than previously available corpora of this type. The dialogues are annotated with dialogue states and system dialogue acts. The dataset [3] was created by the Dialogue Systems Group at University of Cambridge, who also developed the PyDial toolkit used by CityTalk. The specific places mentioned in the dialogues (hotels, restaurants, attractions, hospitals, etc.) are located in the city of Cambridge.

To use MultiWOZ with Rasa, the dataset must be transformed into the required formats. Conversion programs[8] for this task were developed by the Rasa team in order to use MultiWOZ version 2.1 [5] as a benchmark for evaluating Rasa's Transformer Embedding Dialogue (TED) policy [14].

While recognizing that learning to model large-scale multi-domain datasets such as MultiWOZ is among the most challenging tasks for dialogue systems, the Rasa team have noted a number of issues with such datasets [8]. Ambiguous actions are one issue. A system trained on the corpus cannot learn deterministic behaviour, as different actions follow from the same dialogue history in different dialogues. Another issue is that dialogue history does not play a significant role, as knowing only the last user input and its preceding system action achieves the same result as knowing the earlier dialogue history.

## 6 Conclusion and Future Work

After summarizing our earlier work on WikiTalk and CityTalk, we described Rasa open-source conversational AI and how we are using it. We have adapted CityTalk from PyDial to Rasa with encouraging results, and have mentioned some of the differences between PyDial and Rasa.

We also described our use of knowledgebase actions in Rasa and how we have moved from a simple knowledgebase in JSON format to more advanced knowledge graphs in Neo4j graph databases. In ongoing work we aim to enrich the knowledge graphs incrementally by adding further semantic context in the form of taxonomies and other hierarchies of different kinds.

---

[7] MultiWOZ datasets: https://github.com/budzianowski/multiwoz.

[8] MultiWOZ to Rasa conversion: https://github.com/RasaHQ/TED-paper.

In future work we would like to integrate WikiTalk and CityTalk in a single dialogue system. We hope that by using transformer-based dialogue policies in Rasa, which can handle user intents more flexibly, it will be possible to combine both task-oriented dialogues such as hotel and restaurant searches (like CityTalk) and open-domain chat-oriented dialogues that use Wikipedia information (like WikiTalk) within a single integrated dialogue system.

**Acknowledgments.** We thank Kristiina Jokinen (AI Research Center, AIST Tokyo Waterfront) for suggesting the use of Rasa conversational AI.

## References

1. Barrasa, J., Hodler, A.E., Webber, J.: Knowledge Graphs: Data in Context for Responsive Businesses. O'Reilly Media, Newton (2021)
2. Bocklisch, T., Faulkner, J., Pawlowski, N., Nichol, A.: Rasa: open source language understanding and dialogue management (2017). arXiv:1712.05181
3. Budzianowski, P., et al.: MultiWOZ – a large-scale multi-domain wizard-of-oz dataset for task-oriented dialogue modelling (2018). arXiv:1810.00278
4. Bunk, T., Varshneya, D., Vlasov, V., Nichol, A.: DIET: Lightweight language understanding for dialogue systems (2020). arXiv:2004.09936
5. Eric, M., et al.: MultiWOZ 2.1: a consolidated multi-domain dialogue dataset with state corrections and state tracking baselines (2019). arXiv:1907.01669
6. Jokinen, K., Wilcock, G.: Multimodal open-domain conversations with the Nao robot. In: Mariani, J., Rosset, S., Garnier-Rizet, M., Devillers, L. (eds.) Natural Interaction with Robots, Knowbots and Smartphones: Putting Spoken Dialogue Systems into Practice, pp. 213–224. Springer, Cham (2014). https://doi.org/10.1007/978-1-4614-8280-2_19
7. Lala, D., Wilcock, G., Jokinen, K., Kawahara, T.: ERICA and WikiTalk. In: Proceedings of the Twenty-Eighth International Joint Conference on Artificial Intelligence (IJCAI 2019), Macao, China, pp. 6533–6535 (2019). https://www.ijcai.org/proceedings/2019/947
8. Mosig, J.E.M., Vlasov, V., Nichol, A.: Where is the context? – a critique of recent dialogue datasets (2020). arXiv:2004.10473
9. Mrkšić, N., et al.: Multi-domain dialog state tracking using recurrent neural networks. In: Proceedings of the 53rd Annual Meeting of the Association for Computational Linguistics, pp. 794–799. Association for Computational Linguistics, Beijing (2015). http://aclweb.org/anthology/P15-2130
10. Robinson, I., Webber, J., Eifrem, E.: Graph Databases, 2nd edn. O'Reilly Media, Newton (2015)
11. Ultes, S., et al.: PyDial: a multi-domain statistical dialogue system toolkit. In: Proceedings of ACL 2017, System Demonstrations, Vancouver, Canada, pp. 73–78 (2017). https://aclanthology.org/P17-4013/
12. Vaswani, A., et al.: Attention is all you need (2017). arXiv:1706.03762
13. Vlasov, V., Drissner-Schmid, A., Nichol, A.: Few-shot generalization across dialogue tasks (2018). arXiv:1811.11707v1
14. Vlasov, V., Mosig, J.E.M., Nichol, A.: Dialogue transformers (2019). arXiv:1910.00486

15. Wilcock, G.: WikiTalk: a spoken Wikipedia-based open-domain knowledge access system. In: Proceedings of the COLING 2012 Workshop on Question Answering for Complex Domains, Mumbai, India, pp. 57–69 (2012). https://aclanthology.org/W12-6006/
16. Wilcock, G.: Using a deep learning dialogue research toolkit in a multilingual multidomain practical application. In: Proceedings of the Twenty-Seventh International Joint Conference on Artificial Intelligence (IJCAI 2018), Stockholm, Sweden, pp. 5880–5882 (2018). https://www.ijcai.org/proceedings/2018/869
17. Wilcock, G.: CityTalk: robots that talk to tourists and can switch domains during the dialogue. In: D'Haro, L.F., Banchs, R.E., Li, H. (eds.) 9th International Workshop on Spoken Dialogue Systems Technology, pp. 411–417. Springer, Cham (2019). https://doi.org/10.1007/978-981-13-9443-0_37
18. Wilcock, G., Jokinen, K., Yamamoto, S.: What topic do you want to hear about? A bilingual talking robot using English and Japanese Wikipedias. In: Proceedings of COLING 2016, the 26th International Conference on Computational Linguistics: System Demonstrations, Osaka, Japan (2016). https://aclanthology.org/C16-2025/

# Survey of the VR Environment for Deep Learning Model Development

Taisei Naraha(✉), Kouta Akimoto, and Ikuko Eguchi Yairi

Graduate School of Science and Technology, Sophia University, Tokyo, Japan
naraha@yairilab.net

**Abstract.** As deep learning research develops and models become larger and more complex, there are increasing concerns to deep learning about its' low ability of explanations to humans and its black box characteristics. Model visualization research of DL (Deep Learning) has been attracting attention for a decade for a solution to this concern. VR (Virtual Reality) interaction research between humans and models is a practical means of model visualization research with great potential but is still in the early stages. The purpose of our study is to propose new methods for VR technology to contribute to the development of deep learning models by investigating and implementing the visualization technology of deep learning and VR research projects. In this paper, we also report our two experimental results. One is a web survey using demo images of a PC application and a VR goggles application, and the other is an evaluation experiment using VR goggles.

**Keywords:** Human-computer Interaction · Deep learning · Machine learning

## 1 Introduction

Voices that question the low ability of deep learning to explain to humans and the black-boxing are increasing as deep learning research develops and models become larger and more complex. Visualization research of deep learning models is attracting attention as a coping method for this problem [1, 2]. It has been pointed out that VR interaction research between humans and models (hereinafter abbreviated as VR research) has great potential as one of the means, but it is still in the early stages [3]. Therefore, the purpose of this study is to show new possibilities that VR technology contributes to the development of deep learning research by proposing VR research projects for deep learning. This paper proposes a new method of VR technology that contributes to the development of deep learning models by investigating and implementing visualization technology for deep learning and VR research projects. Section 2 reports an overview and examples of VR environments for deep learning, and proposes challenges in applying VR to deep learning visualization research and solutions to these challenges. This paper also reports the results of two experiments. One is a web survey using demo images of the PC application and VR goggles application in Sect. 3, and the other is an evaluation experiment using VR goggles in Sect. 4.


Y. Takama et al. (Eds.): JSAI 2021, AISC 1423, pp. 154–164, 2022.
https://doi.org/10.1007/978-3-030-96451-1_14

# 2 The VR Environment for Deep Learning Model Development

## 2.1 Model Visualization Research of Deep Learning

The development of deep learning visualization technology over the past decade is now being systematically summarized. For example, Choo et al. [4] investigated visual analysis, information visualization, and machine learning perspectives to enable humans to interpret and control deep learning. Yuan et al. [5] investigated each of the machine learning pipelines. The visual analysis technology at each stage is systematically classified according to purpose.

In the pre-modeling process, visualization technology detects anomalous data, highly redundant teacher labels and features, improves the quality of data and features, and analyzes them interactively by the user [6, 7].

The process during model construction includes (1) technology for graphing changes in model output due to parameter changes with the goals of model understanding/model diagnosis/model operation, (2) visualization of convolutional layers to explain the operation of the model, and (3) visual debugging technology for the learning process, etc. [8–10].

Visualization techniques to understand the results of data analysis are essential in the post-modeling process [11, 12].

## 2.2 VR for DL Model Visualization

VR technology is characterized by its high degree of interactivity, which offers clear advantages over conventional visualization approaches, such as improved user comprehension. In a virtual world designed to imitate the physical world, the user is free to maneuver, and with minimal external intervention on the user, the user can achieve a sense of immersion.

Immersion is defined as the feeling that the user is actually in the environment [13]. Even applications that do not resemble reality can be attributed to this phenomenon. Immersion in the senses has benefits in a wide range of fields. Application designs range from entertainment and education to psychotherapy [14]. Researchers believe that VR platforms are more effective than conventional methods for maintaining attention and focus [15]. Such VR environments have also been shown to motivate users to learn deep learning models and positively impact affective and cognitive factors on learning outcomes [16].

Meissler et al. [3] confirmed that visualizing LeNet5 in a VR environment allows users with little knowledge of machine learning to understand the learning process of a learning model with less effort and without being overwhelmed by the amount of information. VanHorn et al. [17] have shown a deep convolutional VR visualization development environment for neural network models and attempted to develop deep learning (DL) models for image classification by biomedical experts in VR. The results showed that the proposed approach could streamline the production and lead to better visual insights even for experienced developers. Applying the classification in the human-centered design of visualization of deep learning presented by Hohman et al. [18], these approaches correspond to "examples of applying VR technology to interact with humans and deep learning through interactive exploration."

### 2.3 Challenges for Visualization Using VR

The issues to be solved when realizing the interaction between humans and deep learning through interactive search using VR technology are summarized below.

1. Evaluation method of usefulness
   When evaluating the usability and usefulness of visualization/analysis methods using VR, it is crucial to introduce the concept of quantification of interpretability, which is the recent research results [19, 20], as an evaluation index. It is also essential that a comparison with desktop applications is necessary as a fundamental matter.
2. High-quality operability
   The proposed method needs to have a user interface that enables stress-free, intuitive, and smooth operation for items of various granularity related to model training and construction.
3. Wide range of customizability
   Since the operations and presentation of information are diverse with various particle sizes, it is important for users to customize the visual structure in the space according to their preference. In particular, customization of parts related to practical model training such as optimization methods and layer properties is an essential issue.
4. Scalability for visual effects and processing
   Visual scalability doesn't overwhelm the user as the model gets deeper and more complex, and system scalability doesn't slow down performance or increase development time when processing the model. Need to be considered.
5. Special support for beginners
   It is easy for novice users to get confused with multiple purposes such as correct operation and understanding of deep learning. It is necessary to devise ways to make it easier for users to understand which problem they are solving.

### 2.4 Solution for Issues of Visualization Using VR

Deep learning and human interaction with VR are fusion research in CHI and artificial intelligence, and it is required to incorporate both perspectives. As a solution to the issues of visualization using VR shown in Sect. 4, the research topics that are important from the viewpoint of CHI are listed below.

- Introduction of new objective and quantitative evaluation criteria such as existing physiological indicators and statistical indicators, and evaluation of their validity
- Introduction of gamification that is especially effective for beginners
- Implementation of an intuitive and scalable model construction environment that emphasizes appearance and operability for intermediate/advanced users and model developers
- Questionnaire-based survey and analysis of awareness of model building process improvement for intermediate/advanced users and model developers
- Introduction of presentation and guide functions that enable selection of better parameter adjustment and optimization methods for practical model training for beginners, intermediate and advanced users

- Investigating the relationship between user satisfaction with system use and interpretability for all users. For interpretability, the proposed framework for quantifying interpretability is applied.
- Investigating the advantages and disadvantages of expressing existing deep learning visualization algorithms in VR for all users
- Survey of user comparison of deep learning visualization application between VR version and desktop version for all users the issues to be solved when realizing the interaction between humans and deep learning through interactive search using VR technology are summarized below.

## 3 Questionnaire Survey for User Needs Analysis

### 3.1 Methodology

In order to identify the need of users for the solutions presented in Sect. 5, we conducted an online survey on training applications for deep learning using VR interaction techniques among 18 people, ranging from beginners to intermediate and advanced users of machine learning. Figure 1 is part of a demo video showing the potential of interactions given to a model in VR space. Figure 2 is an example screenshot of the desktop version of the application. After answering the first question about the user, the survey participant watched the VR video in the browser and operated the downloaded desktop version of the application to compare the two. After the demonstration, the participants rated the four items of operation/spatial arrangement/operation intuition/understanding of visualization regarding the ease of use of the application, and the two items of the impact of the training process on learning/efficiency of model training regarding the effectiveness of the application, using a five-point Likert scale. After that, the participant rated the importance of each of the questions related to the proposed problem-solving items using VR visualization technology, such as gamification in the model training and learning

**Fig. 1.** Data visualization using ArcFace [21]. Display the data after deep learning model training as a group of spheres in VR space.

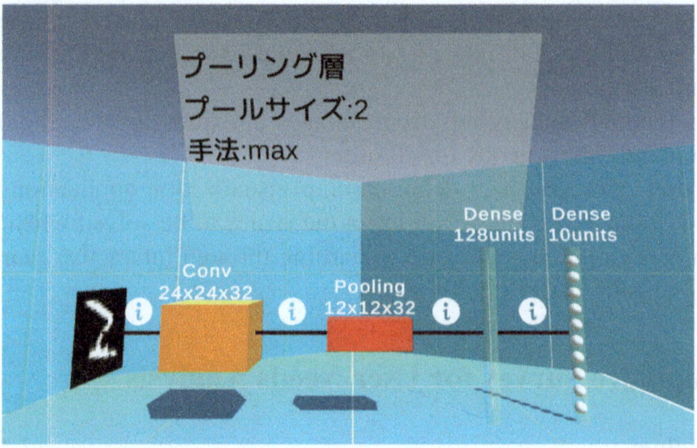


**Fig. 2.** Visualization of network models in desktop applications. The details of the structure and filters of the deep learning model are displayed by operating with the keyboard and mouse.

process and the introduction of functional visualization technology, using a 5-point scale. Our series of online surveys were conducted using a web form.

### 3.2 Result

As shown in Fig. 3, the questionnaire survey results showed that VR applications received higher ratings than desktop applications in all usefulness categories, including operation, spatial arrangement, the intuition of operation, and easy-to-understand visualization. Many respondents said that the simple and intuitive operation in the VR space enhanced the application's attractiveness. As shown in Fig. 4, as for the response tendency regarding effectiveness, VR application was highly evaluated for the learning effectiveness of the training process, while the desktop application was highly evaluated for the efficiency of model training. This result could indicate that the demonstration video was insufficient in presenting the functions necessary for actual model training even if the appearance and operability of the VR application effectively motivated users to learn.

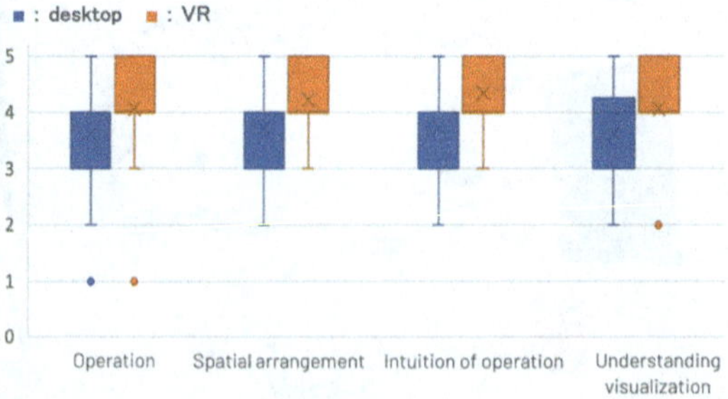


**Fig. 3.** Comparing the ease of use of VR applications and desktop applications. The VR application was highly evaluated in all items of usefulness, such as operation, spatial arrangement, intuition of operation, and easy-to-understand visualization, compared to the desktop application.

In addition, the questionnaire also included questions asking about the importance of "the issues to be solved" in Sect. 4, (1)–(5), while showing more concrete examples. As a result, (2) High-quality operability and (5) Special support for beginners were the highest values, and (1) Evaluation method of usefulness was the lowest value. In the free opinions about VR application development, the functions related to parameter tuning and result presentation for model training and model building were more critical.

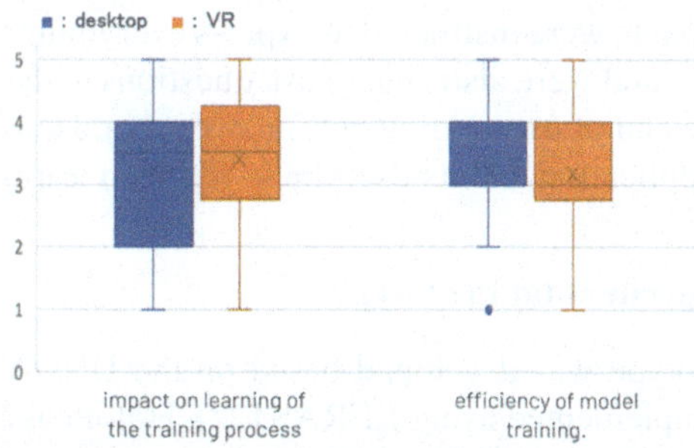


**Fig. 4.** Comparing the effectiveness for deep learning. The VR application was highly evaluated for the effect on learning the training process, while the desktop application was highly evaluated for the efficiency of model training.

## 4 Survey Using VR Goggles

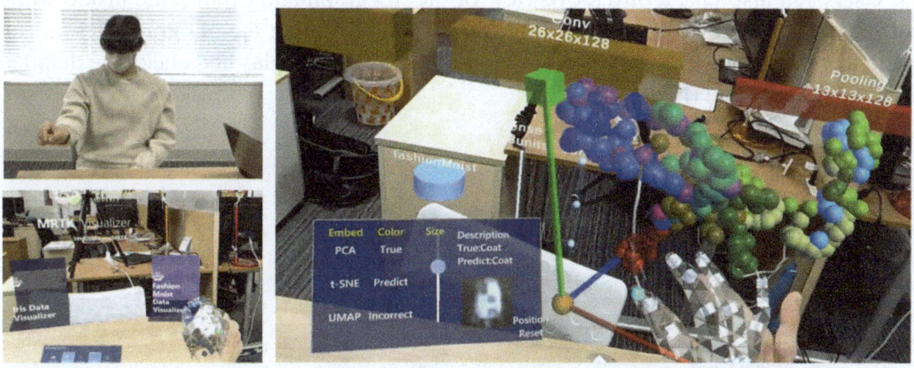


**Fig. 5.** Experiment using VR goggles. Users can add intuitive operations using their own hands recognized by the HoloLens 2 camera.

### 4.1 Methodology

A survey was also conducted on how users who wore VR goggles used the created VR application. Six participants (male 6; age range 21–25 years) took part in the survey. All participants had normal hearing and normal or corrected to normal vision. Participants were students with beginner or intermediate knowledge about deep learning. After receiving an explanation about the purpose of the experiment and the use, the participants wore VR goggles and freely operated the application for an average of 15 min, as shown in Fig. 5. Participants were instructed to express everything they thought verbally during the free operation and were also able to ask questions to the experimenter during free operation. After the end of use, participants answered oral questionnaires regarding the ease of use of the application and its usefulness for deep learning.

### 4.2 Details of the Experimental Prototype

A VR application prototype was developed based on the UNITY platform. The deep learning network was implemented using KERAS [22]. HoloLens2, made by Microsoft, was used as the hardware for development. Users can add intuitive operations such as pushing and picking on deep learning objects using their own hands recognized by the HoloLens 2 built-in camera.

In the AR space, test data is classified by the trained deep learning model. The test data is three-dimensionally compressed, and the output of the intermediate layer with the test data as input is also three-dimensionally compressed. Users can check the 3D graph displaying the data and the 3D model representing the deep learning model in three dimensions.

Users can also take a bird's-eye view of the 3D graph and explore the inside while performing operations such as moving, rotating, and scaling the 3D graph. Users can touch and select the data on the 3D graph one by one and display and check the detailed attributes of the data (Fig. 6).

**Fig. 6.** 3D model and 3D graph of the MLP and CNN. The 3D graph shows the test data and the data output from the intermediate layer with the test data as input, obtained by 3D compression.

The models used in this study are MLP to classify iris datasets (data sets related to iris types and features) shown in Fig. 7, and MLP and CNN to classify fashion MNIST (data set of fashion types and small grayscale images) shown in Fig. 8. Users can specify the data displayed on the 3D graph by selecting each layer on these three deep learning models. In this survey, the MLP input layer and the intermediate layer, the CNN input layer, and the Pooling layer, and the fully connected layer can be selected. In the menu placed next to the model, the dimensional compression method (PCA, t-SNE, and UMA) and the data color-coding method (correct label, predicted label, misclassified data) are displayed as options. The details of the data when the data on the 3D graph is selected are also displayed. As shown in Fig. 9, a user can move and display the menu in space according to the user's request by pointing the palm toward the camera.

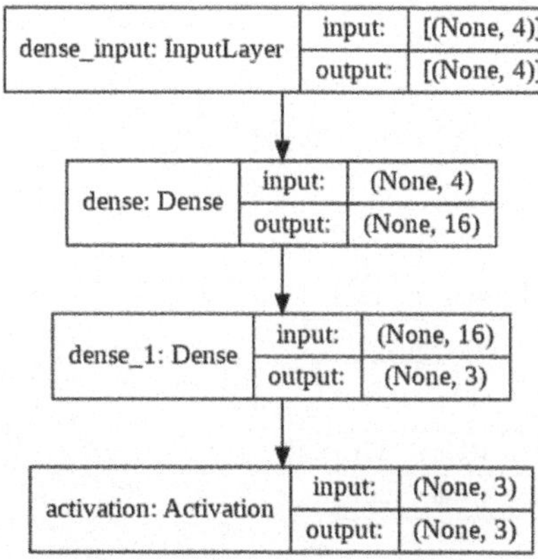


**Fig. 7.** MLP to classify Iris dataset.

### 4.3 Result

All users said they found the application and its purpose interesting. The positive opinions about this application were as follows.

- By comparing the spread and color-coding of the data on the 3D graph between the input layer and the intermediate layer, it was possible to visually imagine how the intermediate layer contributes to the prediction results.
- We were able to infer the tendency of data prediction visually and misclassified data from the search of 3D graphs.
- The VR environment in the application is fun and exciting
- The use for 15 min was enough to get used to the operation.

On the other hand, the following negative opinions were also received.

- The current operation design makes the user feel stressed.
- Current applications lack the ability to give users clear hints for model training.

The following opinions were received as the expected merits of using this application.

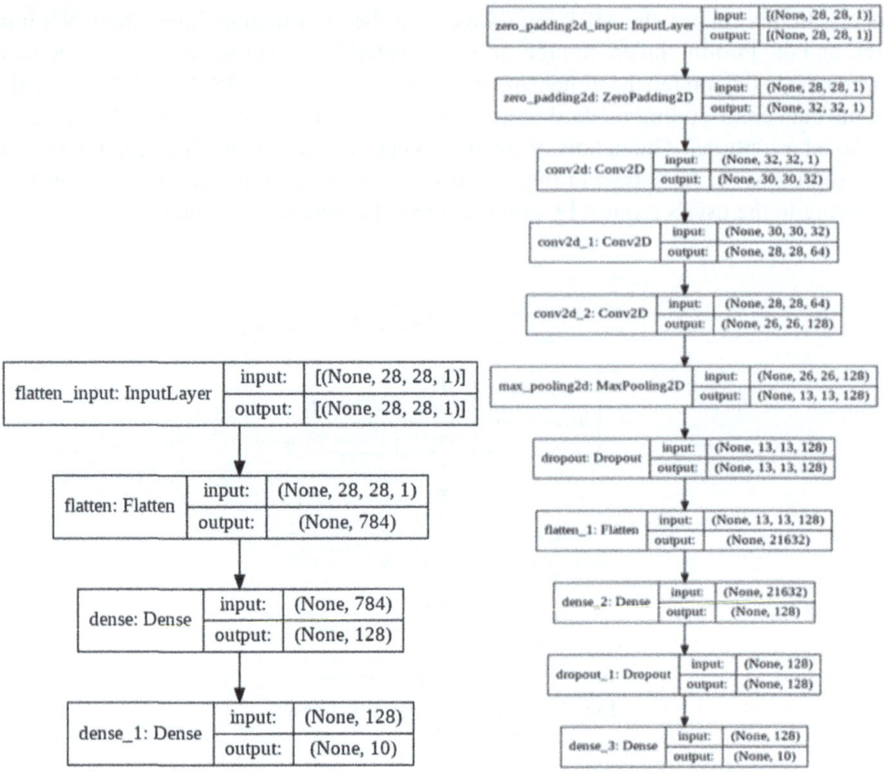


**Fig. 8.** MLP and CNN to classify fashion MNIST.

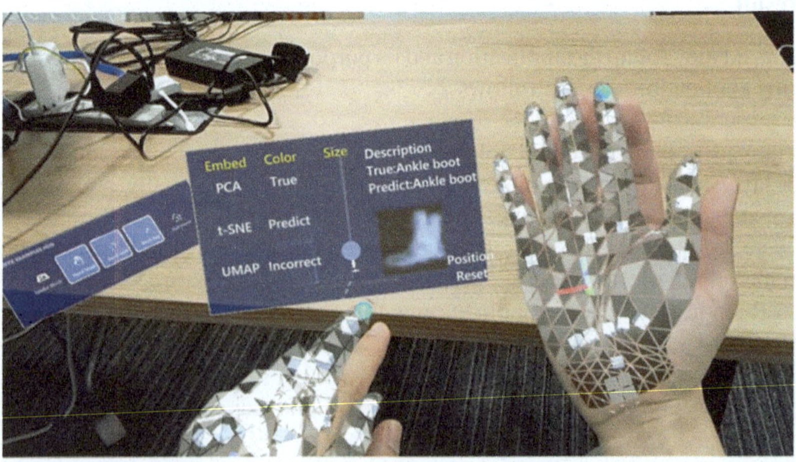


**Fig. 9.** The data display method selection from the menu object. Users can select the dimension compression method and the data color-coding method from the menu and the size of the data on the 3D graph. The details of the data on the 3D graph are displayed on the right side of the menu. The menu can be followed by pointing the palm at the camera, and it can be moved according to the user's request.

- The VR environment has a positive impact on learning motivation and is suitable as a tool for new learners of deep learning to work on deep learning without feeling a mental barrier.
- Effective when checking the input data and its classification results while searching by distributing them widely in 3D space.
- It is more effective when interacting with a large model using ample space, comparing multiple 3D graphs side by side, and searching for 3D graphs from the inside.

The current implementation of our application is still at the concept level. However, through experiments using the application, this paper was able to show that the interaction between deep learning and humans using VR is essentially effective for learning deep learning techniques and improving the efficiency of model training and model construction.

## 5 Conclusion

This paper clarified the challenges for realizing interaction between humans and deep learning using VR by conducting a literature search on visualization technology for deep learning and VR technology for deep learning. The importance of the clarified issues was confirmed through a questionnaire survey comparing the demo video of the VR application and the application running on the browser. The survey of the prototype system using VR goggles was able to show the possibility that VR visualization contributes to deep learning, as well as hints for improving applications and the usefulness that the VR environment gives to users. In the survey, it was easier for the users to observe the object from various angles while walking around the space and changing the viewpoint, rather than moving and rotating the object. In other words, there is a possibility of application development, such as comparing multiple models using a large space and making it possible to understand the characteristics of data by searching the data space. For realization of a practical application, it is important to implement the function to customize the filter and perform model training in real-time and the function to effectively visualize the results so that the user can get direct hints for improving the model.

**Acknowledgement.** We are grateful to all participants of our experiments. This work was supported by a Grant-in-Aid for Scientific Research (B) (No. 20H04476).

## References

1. Liu, S., Wang, X., Liu, M., Zhu, J.: Towards better analysis of machine learning models: a visual analytics perspective. Visual Inform. **1**(1), 48–56 (2017)
2. Zeiler, M., Fergus, R.: Visualizing and understanding convolutional networks. In: Fleet, D., Pajdla, T., Schiele, B., Tuytelaars, T. (eds.) ECCV 2014. LNCS, vol. 8689, pp. 818–833. Springer, Cham (2014). https://doi.org/10.1007/978-3-319-10590-1_53
3. Meissler, N., Wohlan, A., Hochgeschwender, N., Schreiber, A.: Using visualization of convolutional neural networks in virtual reality for machine learning newcomers (2019)

4. Choo, J., Liu, S.: Visual analytics for explainable deep learning. IEEE Comput. Graph. Appl. **38**(4), 84–92 (2018)
5. Yuan, J., Chen, C., Yang, W., Liu, M., Xia, J., Liu, S.: A survey of visual analytics techniques for machine learning. Comput. Visual Media **7**(1), 3–36 (2020). https://doi.org/10.1007/s41095-020-0191-7
6. Chen, C., et al.: Oodanalyzer: interactive analysis of out-of-distribution samples. IEEE Trans. Visual Comput. Graph. **27**(7), 3335–3349 (2020)
7. Liu, S., Chen, C., Lu, Y., Ouyang, F., Wang, B.: An interactive method to improve crowdsourced annotations. IEEE Trans. Visual Comput. Graph. **25**(1), 235–245 (2018)
8. Liu, M., Shi, J., Li, Z., Li, C., Zhu, J., Liu, S.: Towards better analysis of deep convolutional neural networks. IEEE Trans. Visual Comput. Graph. **23**(1), 91–100 (2016)
9. Liu, M., Liu, S., Su, H., Cao, K., Zhu, J.: Analyzing the noise robustness of deep neural networks. In: 2018 IEEE Conference on Visual Analytics Science and Technology (VAST), pp. 60–71. IEEE (2018)
10. Yang, W., Wang, X., Lu, J., Dou, W., Liu, S.: Interactive steering of hierarchical clustering. IEEE Trans. Visual. Comput. Graph. (2020)
11. Jänicke, H., Borgo, R., Mason, J.S., Chen, M.: SoundRiver: semantically-rich sound illustration. In: Computer Graphics Forum, vol. 29, no. 2, pp. 357–366. Blackwell Publishing Ltd., Oxford (2010)
12. Cui, W., et al.: Textflow: towards better understanding of evolving topics in text. IEEE Trans. Visual Comput. Graph. **17**(12), 2412–2421 (2011)
13. Schuemie, M.J., Van Der Straaten, P., Krijn, M., Van Der Mast, C.A.: Research on presence in virtual reality: a survey. Cyberpsychol. Behav. **4**(2), 183–201 (2001)
14. Miller, H.L., Bugnariu, N.L.: Level of immersion in virtual environments impacts the ability to assess and teach social skills in autism spectrum disorder. Cyberpsychol. Behav. Soc. Netw. **19**(4), 246–256 (2016)
15. Sacks, R., Perlman, A., Barak, R.: Construction safety training using immersive virtual reality. Constr. Manag. Econ. **31**(9), 1005–1017 (2013)
16. Witmer, B.G., Singer, M.J.: Measuring presence in virtual environments: a presence questionnaire. Presence **7**(3), 225–240 (1998)
17. VanHorn, K.C., Zinn, M., Cobanoglu, M.C.: Deep learning development environment in virtual reality. arXiv preprint arXiv:1906.05925 (2019)
18. Hohman, F., et al.: Visual analytics in deep learning: an interrogative survey for the next frontiers. IEEE Trans. Visual Comput. Graph. **25**(8), 2674–2693 (2018)
19. Bau, D., Zhou, B., Khosla, A., Oliva, A., Torralba, A.: Network dissection: quantifying interpretability of deep visual representations. In: CVPR (2017)
20. Tsai, C.-Y., Cox, D.D.: Characterizing visual representations within convolutional neural networks: toward a quantitative approach. In: ICML Workshop on Vis for Deep Learning (2016)
21. Deng, J., Guo, J., Xue, N., Zafeiriou, S.: Arcface: additive angular margin loss for deep face recognition. In: Proceedings of the IEEE/CVF Conference on Computer Vision and Pattern Recognition, pp. 4690–4699 (2019)
22. Chollet, F., et al.: Keras. https://keras.io (2015)

# Design and Implementation of a SWI-Prolog-Based Expert System to Diagnose Anxiety Disorder

Kairen Chen and Che-Chern Lin(✉)

Department of Software Engineering and Management, Kaohsiung Normal University, Kaohsiung, ROC
cclin@nknu.edu.tw

**Abstract.** Anxiety disorder is a common mental disorder that affects people's daily life, and causes problems in social activities. Traditionally, doctors have relied on questioning symptoms, observing patients, and even using questionnaires to diagnose and treat their patients, lacking a knowledge framework for detailed and rapid diagnosis. Therefore, an easy-to-use expert system to diagnose anxiety would be very helpful to mental doctors, thus reducing doctors' workload. SWI-Prolog is one of the popular expert system programming languages for building expert systems. Based on Diagnostic and Statistical Manual of Mental Disorders, Fifth Edition (DSM–5) which is a common and useful guidebook for diagnosing mental disorders, this study designed and implemented an expert system for diagnosing anxiety disorders using SWI-Prolog and Java. This study also briefly introduces the basic syntax of SWI-Prolog, and explains how to connect Java and SWI-Prolog. The main advantage of this study is that the couplings of the software modules are low. This leads to high maintainability and scalability.

**Keywords:** SWI-Prolog · Anxiety · Expert systems · DSM-5 · Mental disorder

## 1 Introduction

Anxiety, also known as anxious neurosis, is one of the common mental disorders for human beings. From the viewpoint of mental health, anxiety is considered to be the response of our body to stress. Most anxiety results from nervousness or fear of coming things. For example, we might be nervous or even fearful about tomorrow's examination, job interview, presentation to the public, etc. Nervousness is a common emotion in our daily life. However, if this kind of nervousness has lasted for a certain period (e.g., six months) and such emotion has affected our regular life, it might be an anxiety disorder. Due to the increasing stress of modern society, people might be easy to get anxiety, and such kind of mental disorder might possibly interfere in our daily life.

To diagnose anxiety, a diagnosis guidebook for mental disorders called Diagnostic and Statistical Manual of Mental Disorders, Fifth Edition (DSM–5) [1] has been developed, which indicates useful criteria to diagnose mental disorders. In DSM-5, anxiety has been divided into eleven sub-types. These sub-types are stated as follows [1].


Y. Takama et al. (Eds.): JSAI 2021, AISC 1423, pp. 165–177, 2022.
https://doi.org/10.1007/978-3-030-96451-1_15

1. Medication-induced anxiety disorder.
2. Anxiety disorder due to another medical condition.
3. Specific phobias.
4. Agoraphobia.
5. Panic disorder.
6. Generalized anxiety disorder.
7. Social anxiety disorder.
8. Selective mutism.
9. Separation anxiety disorder.
10. Other specific anxiety disorders.
11. Unspecified anxiety disorder.

Each of the eleven sub-types has different symptoms and treatments. For mental doctors, it is exhausted work to precisely diagnose these sub-types through different symptoms patients might have. An easy-to-use tool to diagnose sub-types of anxiety is therefore necessary for mental doctors. In artificial intelligence, expert systems use well-defined knowledge ontology, called rule base, to mimic the decision-making process of human experts. Essentially, an expert system is a computer program which makes decisions by its rule base. There are basically three main components in an expert system: a rule base, an inference engine, and user interfaces. The rule base in an expert system stores a set of "IF-THEN" rules by which decisions are made via an inference process. The inference engine performs reasoning according to facts (current existing situations) and the rules stored in a rule base. User interfaces provide accessing mechanisms for users to manipulate expert systems. It is very popular to use expert system programing languages to build expert systems. Basically, an expert system programming language provides a built-in inference engine, which saves a lot of programming work to design inference algorithms. With expert system programming languages, expert system developers focus their work simply on how to design rule bases, instead of how to write code to perform reasoning. SWI-Prolog [2] is one of the popular expert system programming languages, which uses a backward reasoning mechanism to conduct inference. It is free software with a lot of supporting materials from the Internet.

## 2 Related Work

Masri and Jani [3] used a hybrid method combining three artificial intelligence approaches to help Malaysian psychology staff to diagnose and treat mental disorders. This hybrid method included rule-based reasoning, fuzzy logic, and a fuzzy genetic algorithm. In this hybrid method, the rules were extracted from human experts of mental health, and the fuzzy logic served to determine the severity levels of mental disorders; finally, the fuzzy genetic algorithm generated appropriate treatment according to patients' budget and heath situations. Hsu and Lin [4] have proposed a conceptual framework for SWI-Prolog-based expert systems to diagnose and treat anxiety using a concept map. This concept map consisted of three parts for different sub-types of anxiety: symptoms, treatments, and side-effects. Hossain et al. [5] have proposed a belief rule-based expert system for diagnosing mental disorders where fuzzy logic was used to present knowledge structure to evaluate the uncertainty of symptoms for mental disorder patients. Praditasari et al. [6] have also implemented an Android-based application to diagnose psychoneurosis and psychosomatic disorder using a backward chaining approach. This Android-based application has been reported that it had high satisfaction in user experience. In addition, Windriyani et al. [7] utilized a forward chaining method to establish an expert system for the diagnosis of mental disorders, where the knowledge base was established according to MINI ICD-10 (Mini-International Neuropsychiatric Interview, the 10th version of the International Classification of Diseases). This forward chaining-based expert system has been reported that it has a 96% diagnosis accuracy where 100 cases were tested. Based on DSM-5, Nunes et al. [8] have proposed a multi-criteria decision model to diagnose different types of mental disorders. The work by Nunes et al. focused on early diagnosis of mental disorders based on qualitative and comparative analyses. Thermolia et al. [9] have presented a framework of remote decision support system to predict crucial changes of bipolar disorder patients. This decision support system could remind mental health staff of the occurrence of critical events, and then suggest suitable treatment for patients.

## 3 Syntax of SWI-Prolog

In SWI-Prolog, four kinds of basic elements are utilized: atoms, numbers, variables, and complex terms [2]. An atom beginning with a lower-case letter simply indicates a particular item. A number directly expresses a particular numerical value, either an integer or a floating-point number. A variable beginning with an upper-case letter is a space to store any possible atoms as long as conditions are met. Variables are particularly useful when an inference process is conducted by SWI-Prolog's built-in inference engine. A complex term is a structure-like item possibly consisting of atoms, numbers, and variables. In most cases, a complex term is a functor-like structure followed by a sequence of arguments.

In SWI-Prolog, three types of syntax are commonly used: facts, rules, and queries. We explain them in the following subsections.

### 3.1 Fact

A fact simply describes a current condition for further reasoning. In SWI-Prolog, the syntax of facts is defined as follows.

fact_name (arg_list):- true (or false).

where fact_name is the name of this fact, arg_list is the list of arguments, and true (or false) indicates whether this fact exists (true) or not (false). For example, the following fact describes the fact that "piano is a music instrument".

music_instrument(piano):- true.

### 3.2 Rule

Basically, a rule is an "IF-THEN" structured syntax, as shown as follows.

rule_name(arg_list):- condition 1, condition 2, …, condition *n*.

where rule_name is the name of this rule, arg_list is the argument list, and condition 1 to condition *n* are the necessary conditions for this rule. In a rule, the rule_name is considered as the "THEN" part of this rule, and condition 1 to condition *n* are considered as the "IF" part of this rule. These conditions are connected by commas which function like logical "AND". For example, the following rule is explained as "if a food is soft and this food is non-spicy, then peter likes this food".

likes(peter, Food): soft(Food), non_spicy (Food).

It is important to note that semicolons can be used to connect the conditions in a rule, if these conditions are connected by logic "OR". In the above rule, a word beginning with a lower-case letter is an atom (item) representing this item itself. On the other hand, a word beginning with an upper-case letter is a variable representing any possible items as long as conditions are met. For example, in the above rule, "peter" is an atom representing peter himself. However, "Food" is a variable representing any kind of food as long as it is soft and non-spicy.

### 3.3 Query

A query is a way to get reasoning results based on a set of rules fired by facts. It is very common to use variables in a query to get desired reasoning results. The results of a query are basically obtained by the inference procedure conducted by the built-in inference engine of SWI-Prolog.

### 3.4 Example

Consider a simple rule base consisting of five facts and one rule as follows.

japanese(yakiniku) :- true.
japanese(sashimi) :- true.
japanese(lamen) :- true.
non_raw(yakiniku) :- true.
non_raw(lamen) :- true.
likes(peter, Food) :- japanese(Food), non_raw(Food).

<u>**Case 1:**</u> Query whose answer is either true or false.
If we key in a query "likes(peter, lamen).", the answer is true. On the other hand, if we key in a query "likes(peter, sashimi).", the answer is false.

<u>**Case 2:**</u> Query whose answers are multiple items.
If we key in a query "likes(peter, X).", SWI-Prolog will return the following results.

X = yakiniku;
X = lamen.

## 4 System Design and Implementation

In this section, we present the framework of this study, illustrate Java implementation with code fragments, and describe how to design the rule bases for the 11 sub-types of anxiety [10]. Finally, we show the main GUI of this study, and briefly explain the manipulation of the main GUI [10].

### 4.1 Framework

The user interface of SWI-Prolog is a command-line-based console. It is not suitable for modern computer applications. Besides, it is very difficult for SWI-Prolog to perform traditional flow control (e.g., for/while looping, if/else branches, etc.) and mathematical manipulation. Therefore, people often use a Java Archive file called JPL (jpl.jar) to connect Java and SWI-Prolog [11]. Strategically, we use SWI-Prolog to perform logical reasoning; on the other hand, we utilize Java to generate graphical user interfaces (GUIs), do mathematical manipulation, and perform business flow control. Figure 1 shows the framework of our system [4].

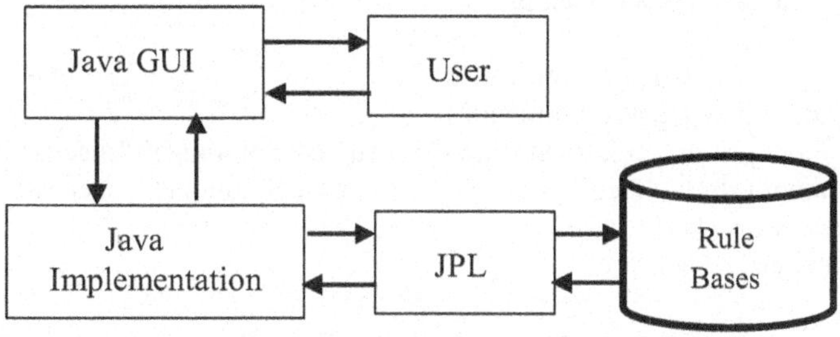


**Fig. 1.** Framework of our system [4]

### 4.2 Java Implementation with JPL

SWI-Prolog is powerful in performing logical inference based on pre-defined rules. However, it uses a command-line-based console for users to key in queries and to display inference results. This is not suitable for modern computer applications. Besides, in SWI-Prolog it is very difficult to perform flow control and to do mathematical manipulation. A bridging interface between SWI-Prolog and a traditional programming language (e.g., Java) is needed to help SWI-Prolog developers to build their GUIs. Moreover, Java can be practically used to perform business flow control and to do necessary mathematical manipulation.

JPL is a bidirectional interface between Java and Prolog, containing a set of Java class and C functions [11]. With JPL, SWI-Prolog developers cab fast and easily build their GUI-based applications. Next, we will describe the procedure of using JPL and provide some code fragments regarding the interactions between Java and SWI-Prolog.

The procedure of using JPL is illustrated as follows.

1. Link rule bases.
2. Send queries to SWI-Prolog based on the facts collected from GUIs.
3. Get inference results from SWI-Prolog.
4. Display the results on GUIs.

Here, we provide some code fragments of Java for the use of JPL.

**Linking rule bases**

In SWI-Prolog, the file extension of rule bases is "pl". It is very common to use multiple rule bases for an application. It is also very common to store a set of facts in a rule base. Assume that the rules are stored in a rule base called rule.pl, and the facts are stored in a rule base called fact.pl. The following code fragment shows us how to link the two rule bases to the system. Class Query shown in this code fragment is a class for doing queries, which is defined in JPL. Therefore, it is necessary to import JPL before performing the following code fragment. (i.e., import org.jpl7.*;).

```
String commend1 = "consult(rule.pl).";
String commend2 = "consult(fact.pl).";
try {
    Query q1 = new Query(commend1);
    Query q2 = new Query(commend2);
    System.out.println("rules connected? " + (q1.hasSolution()? "success" : "fail"));
    System.out.println("facts connected? " + (q2.hasSolution()? "success" : "fail") );
} catch (Exception e1) {
    e.printStackTrace();
}
```

**Sending a query whose answer is either true or false (Case 1 in Sect.** 3.4)

Consider the Case 1 example in Sect. 3.4. If we send a query "likes(peter, lamen).", the answer is true. This query can be implemented by the following code fragment.

```
String command = "likes(peter, lamen).";
Query q = new Query()
boolean answer = q.hasSolution();
System.out.println("The answer is " + answer);
```

**Sending a query whose answers are multiple items (Case 2 in Sect.** 3.4)

This is a more complicated example, since Java uses class Map to handle multiple items. Consider the Case 2 example in Sect. 3.4. We want to send a query "likes(peter, X)." to SWI-Prolog. This query can be implemented by the following code fragment where class Map is a class defined in java.util.Map.

```
String command = "likes(peter, X).";
Query q = new Query(command);
while ( q.hasMoreSolutions() ){
   Map<String, Term> map= q.nextSolution();
     result += map.get("X") + ",";
   }
```

Besides, since the inference results obtained from SWI-Prolog contain some un-expected characters such as ‘[’, ‘]’, ‘\’, ‘(’, ‘)’, etc., it is necessary to remove these un-expected characters for better display of inference results.

### 4.3 Design of Rule Bases

Figure 2 is the logical architecture of the rule bases. Basically, there are two main rule bases in our system: symptoms and anxiety. The functions of the two rule bases are briefly described as follows.

1. Rule Base Symptom: This rules base stores the symptoms a patient might have. These symptoms are collected from GUIs.
2. Rule Base Anxiety: This rule base stores rules to diagnose the 11 sub-types of anxiety. This rule base is established according to the DSM-5 guidebook.

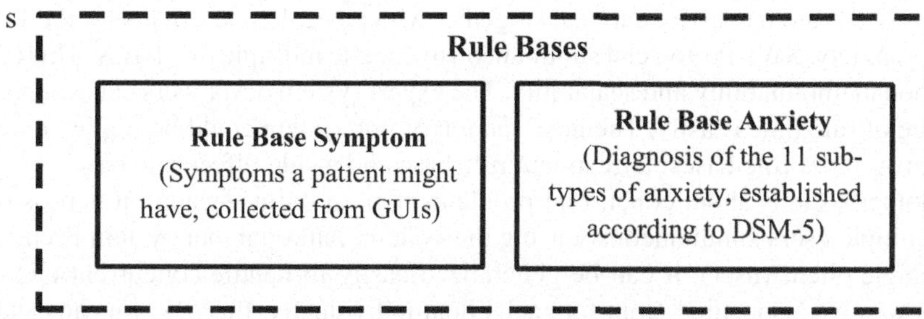


**Fig. 2.** Logic architecture of the rule bases

**<u>Examples</u>**

The following rules are the diagnosis rules for social anxiety disorder, which is one of the 11 sub-types of anxiety.

```
anxiety_range :- anxiety_feeling.
specific_anxiety_range :- anxiety_range, specific_range.
social_range :- specific_anxiety_range, social_scene.
social_anxiety_disorder:- social_range, for_at_least_six_months.
```

The rules and initial facts for this study are shown in Appendixes. It is important to note that all of the initial rules are set to be false at the very beginning. Mental doctors can change the status of initial facts according to patient’s symptoms.

### 4.4 User Interfaces

Figure 3 shows the main GUI of this study. In this figure, a mental doctor can select possible symptoms a patient might have. After confirming the symptoms, the mental doctor then clicks the "submit" button. The possible anxiety disorders associated with these symptoms will be displayed at the bottom of this GUI.

## 5 Discussion

This study proposes an expert system to diagnose anxiety using SWI-Prolog and Java. There are basically two parts in the proposed system: rule bases and GUIs. The rule bases were simply and straightforwardly implemented by translating the guidelines of DSM-5 to SWI-Prolog syntax. The domain knowledge (DSM-5 diagnosis guidelines) could be easily organized in well-structured rules. The GUIs implemented by Java served to collect patients' symptoms and to display the inference results. JPL was used as a bridge to connect the rule bases and the GUIs. The main advantage of the proposed system is that the software coupling between the two parts (the rule bases and the GUIs) is very low. Therefore, the maintainability and scalability of the proposed system are high. Rule base designers only pay their attention to how to develop SWI-Prolog-based rules according to knowledge structure of problem domains. On the other hand, GUIs designers focus only on how to design their GUIs using Java.

The proposed system diagnoses anxiety was based on DSM-5. It is possible to extend the proposed system to give treatments together with possible side-effects for the 11 subtypes of anxiety. SWI-Prolog can simultaneously handle multiple rule bases. This causes very good maintainability and scalability. The expert system developers can extend their structure of rule bases easily. The next version of our system could be implemented by adding two more rule bases: a treatment rule base and a side-effect rule base.

Another issue is about concurrent problems. For an online system, it is possible to have multiple users simultaneously using the system. Although our system is currently for a single client (user), it can be generalized easily to handle concurrent clients as long as we give a specified atom for each client individually. Besides, we can establish a dedicated rule base to individually store the symptoms for each of concurrent clients. For example, consider two clients, peter and tom. We can use two individual rule bases to store the symptoms for the two clients, respectively. Examples of the two rule bases are illustrated as follows.

**Rule base storing the symptoms of peter:**

```
caused_by_medicine(peter) :- false.
caused_by_diseaces(peter) :- false.
panic_feeling(peter) :- true.
......
... more symptoms ...
```

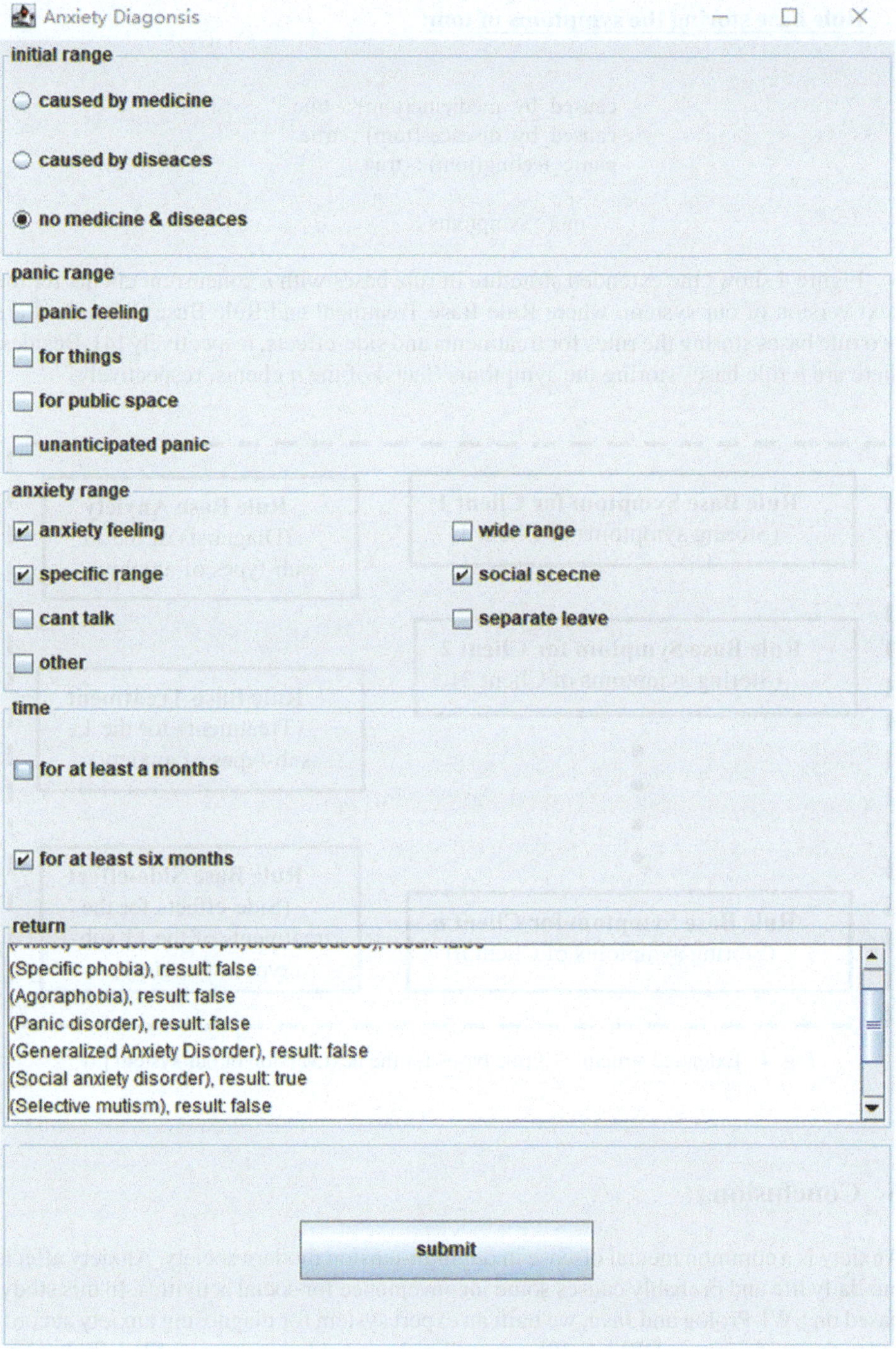


**Fig. 3.** Main GUI of this study

**Rule base storing the symptoms of tom:**

```
caused_by_medicine(tom) :- true.
caused_by_diseaces(tom) :- true.
panic_feeling(tom) :- true.
......
... more symptoms ...
```

Figure 4 shows the extended structure of rule bases with *n* concurrent clients for the next version of our system, where Rule Base Treatment and Rule Base Side-effect are two rule bases storing the rules for treatments and side-effects, respectively [4]. Besides, there are *n* rule bases storing the symptoms (facts) of the *n* clients, respectively.

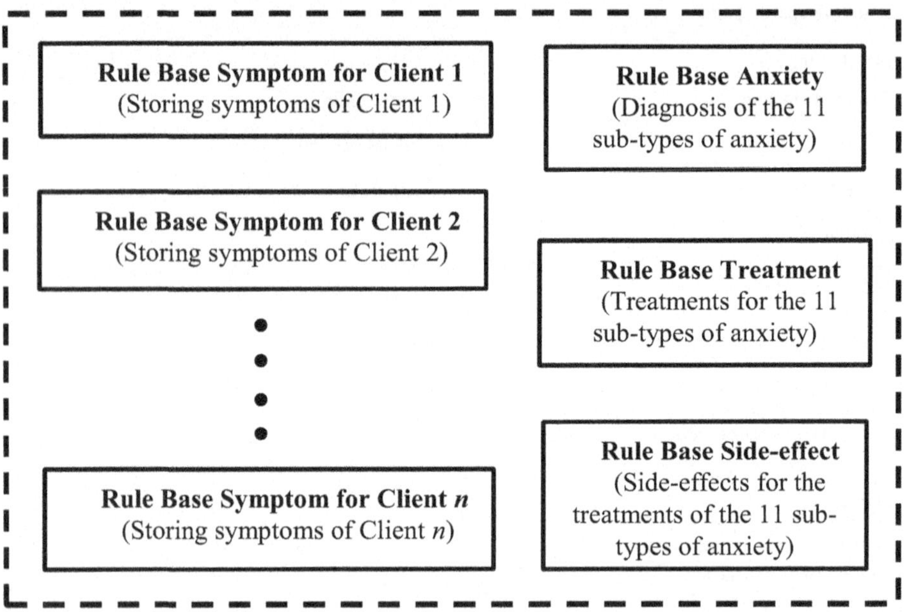


**Fig. 4.** Extended structure of rule bases for the next version of our system [4]

## 6 Conclusion

Anxiety is a common mental disease in our high-tension modern society. Anxiety affects our daily life and probably causes some inconvenience for social activities. In this study, based on SWI-Prolog and Java, we built an expert system for diagnosing anxiety according to the guidelines in DSM-5. JPL was utilized as a bridge to connect SWI-Prolog and Java. We briefly introduced the syntax of SWI-Prolog, and then proposed the framework of this study. We also demonstrated how to use JPL by providing code fragments of

Java, and explained how to design our rule bases. Besides, the main GUI of this study and its manipulation were also illustrated. In the discussion section, the issue related to the maintainability and scalability of the proposed system was discussed. In addition, the concurrent problem issue for multiple users was also considered. It can be simply solved by SWI-Prolog syntax. One might further extend our framework to a web-based application which can be concurrently used by different mental doctors, giving them personalized diagnosis environments according to their individual needs. This might be a novel point of this study.

The advantages of using SWI-Prolog for diagnosing anxiety are stated as follows.

- In the proposed system, the overall software coupling between the Java modules and the rule bases is low. This makes the proposed system easy-to-maintain.
- SWI-Prolog has a built-in inference engine to perform reasoning. Therefore the system developers only pay their attention to design rule bases and GUIs. There is no need to design an inference algorithm.
- Rule bases are straightforward and easy-to-maintain. They could be simply established by translating the guidelines of DSM-5 to SWI-Prolog syntax.
- SWI-Prolog provides a powerful library to conduct more complicated reasoning.
- JPL can be used to connect Java, making applications more powerful in GUI design, mathematical computation, and business flow control.
- Concurrent problems can be easily handled as long as we give each of concurrent users a specified atom and a dedicated rule base.

As for the directions of future studies, the following issues could be considered.

- Rule bases for the treatments and the side-effects of the 11 sub-types of anxiety could be added to the next version of our system.
- One might consider establishing more complete expert systems to diagnose and treat some other mental disorders (e.g., depression, post-traumatic stress disorder, etc.) based on the guidelines illustrated in the DSM-5 guidebook.

## Appendixes: Rules and Initial Facts

### A.1 Rules

medication_induced_anxiety_disorder:- cause_by_medicine.

anxiety_another_body_diseace:- cause_by_diseaces.

specific_panic_range:- panic,for_at_least_six_months.

specific_phobia:- specific_panic_range,for_things.

Agoraphobia:- specific_panic_range,for_public_space.

un_specific_panic_range:- panic,for_at_least_a_months.

panic_disorder:- un_specific_panic_range,unanticipated_panic.

```
anxiety_range:- anxiety.
wide_anxiety_range:- anxiety_range,wide_range.
specific_anxiety_range:- anxiety_range,specific_range.
generalized_anxiety_disorder:- wide_anxiety_range,for_at_least_six_months.
social_range:- specific_anxiety_range,social_scene.
social_phobia:- social_range,for_at_least_six_months.
selective_mutism:- social_range,cant_talk.
separation_anxiety_disorder:- specific_anxiety_range,separate_leave.
other_specified_anxiety_disorder:- anxiety_range,other.
unspecified_anxiety_disorder:- other.
```

## A.2 Initial Facts

```
caused_by_medicine:- false.
caused_by_diseaces:- false.
panic_feeling:- false.
for_at_least_six_months:- false.
for_at_least_a_months:- false.
for_things:- false.
for_public_space:- false.
unanticipated_panic:- false.
anxiety_feeling:- false.
wide_range:- false.
specific_range:- false.
social_scene:- false.
cant_talk:- false.
separate_leave:- false.
other:- false.
```

## References

1. American Psychiatric Association, Diagnostic and statistical manual of mental disorders, fifth edition (DSM-5) (2013)
2. SWI-Prolog Homepage (2021). https://www.swi-prolog.org. Accessed 06 Sept 2021
3. Masri, R.Y., Jani, H.M.: Employing artificial intelligence techniques in mental health diagnostic expert system. In: 2012 International Conference on Computer & Information Science (ICCIS), pp. 495–499 (2012)
4. Hsu, C.-C., Lin, C.-C.: Framework and conceptual design of rule base for building SWI-Prolog-based expert systems to diagnose and treat anxiety. In: 2020 International Conference on Pervasive Artificial Intelligence (ICPAI), pp. 54–57 (2020)

5. Hossain, M.S., Monrat, A.A., Hasan, M., Karim, R., Bhuiyan, T.A., Khalid, M.S.: A belief rule-based expert system to assess mental disorder under uncertainty. In: 5th International Conference on Informatics, Electronics and Vision (ICIEV), Dhaka, Bangladesh, pp. 1089–1094 (2016)
6. Praditasari, W.A.A., Novianti E., Kholis, I., Andriyusadi, R.: Modified backward chaining android application to diagnose psychoneurosis and psychosomatic disorder. In: 6th International Conference on Electrical Engineering, Computer Science and Informatics (EECSI), pp. 147–152 (2019)
7. Windriyani, P., Kom, S., Wiharto, S.T., Kom, M., Sihwi, S.W., Kom, S.: Expert system for detecting mental disorder with forward chaining method. In: International Conference on ICT for Smart Society, Jakarta, Indonesia, pp. 1–7 (2013)
8. Nunes, L.C., Pinheiro, P.R., Pinheiro, M.C.D., Filho, M.S., Nunes, R.E.C., Pinheiro, P.G.C.D.: Automatic detection and diagnosis of neurologic diseases. IEEE Access **7**, 29924–29941 (2019)
9. Thermolia, C.H., Bei, E.S., Petrakis, E.G.M.: TeleCare of mental disorders by applying semantic web technology. In: 13th IEEE International Conference on BioInformatics and BioEngineering, Chania, Greece, pp. 1–4 (2013)
10. Chen, K.: A SWI-Prolog-based expert system to diagnose anxiety with treatment recommendation. Master thesis, Kaohsiung Normal University (2021)
11. JPL Homepage. https://jpl7.org. Accessed 10 Sept 2021

# RandECG: Data Augmentation for Deep Neural Network Based ECG Classification

Naoki Nonaka and Jun Seita(✉)

Advanced Data Science Project, RIKEN Information R&D and Strategy Headquarters, Tokyo, Japan
{naoki.nonaka,jun.seita}@riken.jp

**Abstract.** In the medical field, it is quite expensive to obtain labeled data that are essential to train deep neural networks (DNNs). One way to tackle this issue is to apply data augmentation, a technique to improve classification accuracy by increasing diversity of data through random but realistic transformations. Data augmentation have shown promising results in visual domain, however, transformations applied to image data cannot be directly applied to ECG data. Here we propose RandECG, a data augmentation method tailored for electrocardiogram (ECG) data classification with deep neural networks (DNNs). We explored various transformation methods and selected suitable transformations for ECG. We tested efficacy of RandECG on two different datasets, and found that the classification accuracy of atrial fibrillation can be improved up to 3.51%, without changing an architecture of DNNs.

**Keywords:** ECG · Data augmentation · DNN

## 1 Introduction

Electrocardiogram (ECG) is a widely used device to monitor hearts' electronic activities and is crucial to diagnose cardiovascular diseases. In practice, to diagnose abnormal heart activities, cardiologists review ECG signals which requires a labor-intensive process putting a large burden on the cardiologist. Thus, to alleviate these labor intensive processes, an automatic detection system of abnormal ECG to assist physicians are developed.

In order to automatically detect abnormal ECG, various machine learning approaches have been proposed. Among these approaches, several works have shown deep neural networks (DNNs) can detect irregular ECGs without the needs of feature engineering [2,18]. DNNs have driven substantial advances and demonstrated dramatic improvement of state of the art in tasks like image recognition, machine translation and speech recognition [3,11,17,20,29]. However, DNN models require large amounts of training data, which are often expensive, especially in the medical field.

One way to alleviate the problem of requiring large amounts of training data is data augmentation. Data augmentation is a technique to improve classification


Y. Takama et al. (Eds.): JSAI 2021, AISC 1423, pp. 178–189, 2022.
https://doi.org/10.1007/978-3-030-96451-1_16

accuracy, often applied in domains such as image classification and text classification. Data augmentation increases the amount and diversity of training data by applying transformations based on augmentation operations. For instance, in image domain, augmentation operations include flipping image, random rotation, or random masking [12] and in neural machine translation back-translation is used as an augmentation [27]. Yet, augmentation operations applied to image data can only be applied to image data and not to other modalities of data such as ECGs.

Thus, in this work we explored data augmentation operations tailored for ECG data and propose RandECG which utilizes random combination of transformations to obtain suitable diversity for ECG classification. We first explore ECG data and search for patterns of transformations which appear in ECG data. Subsequently, based on observed transformations, we list several ECG augmentation operations. Following the success of [10] in image classification tasks, we propose RandECG which randomly combines these listed operations. We show by introducing RandECG, we can improve classification of atrial fibrillation with single lead ECG data up to 3.51%, without changing the architecture of classification DNN.

## 2 Related work

Wide variety of research fields have experienced advances with DNNs, including the field of ECG classification, with its ability to automatically extract features. Several researches with end-to-end DNN applied to ECG classification related tasks have been conducted. [18] has shown a cardiologist level classification accuracy with their data using a 34 layers DNN. Also, [2] applied DNN to 12 lead ECGs and showed the model can detect patients who previously had atrial fibrillation. In [25], DNN was applied to detect hypoglycemic events from ECGs. Approaches with DNNs have shown promising results, however, DNN models require large amounts of data, for instance [18] collected 64, 121 labeled data from 29, 163 patients, [2] collected 649, 931 normal sinus rhythms from 180, 922 patients. This requirement for large sized training data often hinders the application of DNN models, especially in the medical field, where acquisition of training data is often expensive.

To mitigate the requirements for large sized training data, there are two major augmentation approaches: one is to train a generative model from the given data and the other is to increase the diversity of data by applying augmentation operations to the data. Approaches using generative models, Generative Adversarial Nets (GANs) [16] are often adopted as a generator of additional data [5,26]. In the context of ECG data, several works have applied GANs to generate ECG data for augmentation [13,14,19]. The advantage of generative models is unlimited number of data generation, nevertheless training deep generative models such as GANs is often difficult with small amounts of training data.

On the other hand, the method of applying augmentation operation to data is simpler than the model with generators since it only performs simple transformations to the data. Recent works have shown image classification accuracy can be improved by combining multiple augmentation operations [9,10]. However, augmentation operations applied to image data can only be applied to image data and not to other modalities such as ECGs. As an augmentation operation for ECG data, [6] enhances DNNs for detecting atrial fibrillation by repeating ECG episodes. In this study, we investigate multiple augmentation operations for ECGs based on transformations appearing in ECG dataset and show that they can improve the classification accuracy by randomly combining them.

## 3 RandECG

Various augmentation methods have been successfully applied to train DNN models in the vision [9,10,12], speech [24] and machine translation [27]. However, these augmentation approaches cannot be directly applied to ECG data.

Thus, in this work we explored data augmentation operations suitable for ECG data. To discover augmentation operations for ECG data, we first explored ECG data and searched for patterns of transformations which appear in ECG data. We explored the PhysioNet/Computing in Cardiology Challenge 2017 (PhysioNet17) dataset [8,15] and observed flipping, difference of amplitudes, partial disturbance shaped similar to sine curve and fluctuation of base lines. Based on these observations, we listed augmentation operations for ECG data. Additionally, we listed augmentation operations such as cutout [12] based on success in image classification tasks and application of finite impulse response (FIR) filter which is sometimes applied as preprocessing of ECG data. The full list of augmentation operations are shown on Table 1.

Subsequently, we determined the range of magnitude and frequency of each augmentation operation by preliminary experiment. As a preliminary experiment we trained model with single augmentation operation applied respectively, using PhysioNet17 dataset. For sine, square pulse, partial sine and partial square pulse, we first determined range of frequency. Frequency range was set as a constant multiple of the frequency of the input data. Afterwards, we determined the magnitude range of each augmentation operation. Based on the results of the preliminary experiment, we determined values of magnitude for each augmentation operation. Selected ranges of magnitude were double or half of best scored results of preliminary experiments.

Recently, [10] reported that random combination of augmentation operations can improve classification accuracy of image data with only two additional hyper parameters. Following the success of [10] in image classification tasks, we propose RandECG which randomly combines augmentation operations for ECG data shown in Table 1. The additional hyper parameters of augmentation is magnitude of each augmentation, $M$ and number of augmentation to apply in each batch, $N$.

**Table 1.** List of augmentation operations tailored for ECG data

| Operation name | Description | Range of Magnitudes |
|---|---|---|
| Identity | Apply no augmentation operation. | – |
| Scale | Randomly scale input signal by random value between *magnitude*. | $[0.25, 4]$ |
| Flip | Randomly flip input signal with probability of *magnitude*. | $[0, 1.0]$ |
| Drop | Randomly mask input signal with probability of *magnitude*. | $[0, 0.4]$ |
| Shift | Randomly shift input signal by *magnitude* × *signal length*. (Direction is randomly chosen by equal probability.) | $[0, 1.0]$ |
| Cutout | Randomly cutout input signal with width of *magnitude* × *signal length*. | $[0, 0.4]$ |
| Sine | Add sine curve with random frequency and amplitude of *magnitude* (Frequency ratio range of $[0.001, 0.02]$ ). | $[0, 1.0]$ |
| Square pulse | Add square pulse with random frequency and amplitude of *magnitude* (Frequency ratio range of $[0.001, 0.1]$). | $[0, 0.02]$ |
| Gaussian noise | Add Gaussian noise multiplied by *magnitude*. | $[0, 0.02]$ |
| Sine partial | Add sine curve of random frequency with width of *magnitude* × *signal length* (Frequency ratio range of $[0.1, 1.0]$). | $[0, 0.1]$ |
| Square pulse partial | Add square pulse of random frequency with width of *magnitude* × *signal length* (Frequency ratio range of $[0.02, 1.0]$). | $[0, 1.0]$ |
| Gaussian noise partial | Add Gaussian noise with width of *magnitude* × *signal length*. | $[0, 0.1]$ |
| FIR lowpass | Apply finite impulse response (FIR) low pass filter with filter value of *magnitude* × *ECG frequency*. | [0.1, 0.4999] |

## 4 Experiment

To examine the efficacy of augmentation operation listed on Table 1 and RandECG, we conducted ECG classification with two different datasets using DNN model. We compare models trained without augmentations, fixed single augmentation applied, model trained with all augmentations always applied and RandECG. We first describe two datasets used for the experiments, and subsequently explain evaluation criteria and model used to classify ECG data.

### 4.1 Dataset

**PhysioNet/Computing in Cardiology Challenge 2017 Dataset.** The PhysioNet/Computing in Cardiology Challenge 2017 (PhysioNet17) dataset consists of ECG data collected by AliveCor devices. We conducted an experiment using the public dataset of the PhysioNet17 dataset, which consist of 8,528 samples. Each sample in the data set is assigned one of four class labels: Normal, atrial fibrillation (AF), other rhythm, and too noisy.

We applied three preprocessing steps to the original PhysioNet17 dataset, namely scaling, down sampling and sequence length alignment. We first scaled the data by dividing the raw ECG signals by absolute value of dynamic range value 500 and subsequently down sampled to 50 Hz frequency. Finally, in order to align the length of samples, we added zero vectors to the head of each sample until the length of vector aligns with the longest sample contained in the dataset. As a result of the processing procedure, all the samples contained in PhysioNet17 dataset have a length of 3,050 dimension.

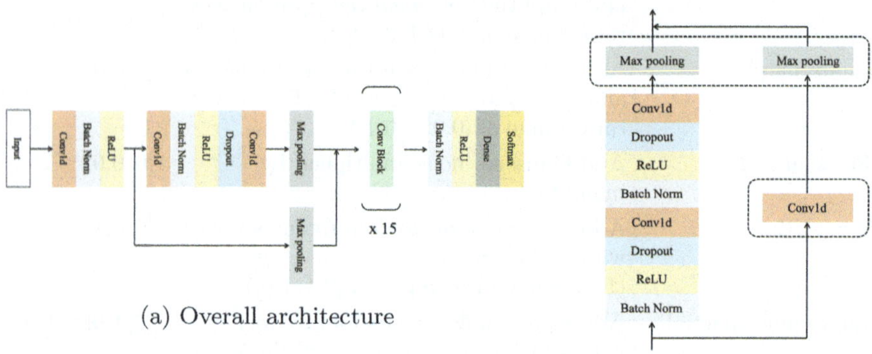


(a) Overall architecture

(b) ConvBlock

**Fig. 1.** Model architecture of ECG classification model. Fig. (a) shows overall architecture and Convblock shown in Fig. (b) is stacked 15 times in the model.

**The China Physiological Signal Challenge 2018 Dataset.** The China Physiological Signal Challenge 2018 (CPSC18) dataset is a dataset containing 12-lead ECGs collected in the intensive care unit. The dataset has nine labels, including normal and atrial fibrillation, with a maximum of three labels assigned to each sample. We used publicly available 6,877 data in our experiments.

In order to treat the data in the same way as the PhysioNet17 dataset, we preprocessed samples in CPSC18 dataset. Applied preprocessing steps were scaling, down sampling and sequence length alignment, followed by label integration and lead I ECG extraction. First, we scaled the samples based on the maximum and minimum values of each collected sample and subsequently the frequency was dropped from 500 Hz to 50 Hz. Subsequently, zero padding was performed to match the length of the ECG data with a maximum length of 80 seconds. Label data were reassigned to align with PhysioNet17 dataset, with following the procedure.

- Normal: assigned to samples originally labeled as normal in CPSC18 dataset.
- AF: assigned to samples originally labeled as AF in CPSC18 dataset, regardless of the other labels.
- Other rhythm: All samples that do not fall under the above two criteria.

Finally, to align the number of leads, we extracted lead I ECG signals from each sample.

### 4.2 Evaluation criteria

Following the evaluation criteria for PhysioNet/Computing in Cardiology Challenge 2017 [8], we calculate F1 score for 3 classes (Normal, AF and Other rhythms). We denote F1score of the model for class $c$ by $F1score_c$, calculated by following equation.

$$F1score_c = \frac{2 \times N_{cc}}{N_{cx} + N_{xc}} \tag{1}$$

where, $c = 1, 2, 3$ for Normal, AF and Other rhythm class respectively. $N_{cc}$ correspond to a number of samples with predicted class of $c$ and reference class of $c$. $N_{cx}$ and $N_{xc}$ corresponds to a number of samples with reference class of $c$ and predicted class of $c$ respectively. The final score of the model was calculated by averaging F1score of three classes.

$$score = \frac{F1score_1 + F1score_2 + F1score_3}{3} \tag{2}$$

Both PhysioNet17 dataset and CPSC18 dataset were split into train, validation and test dataset with size ratio of 0.6, 0.2 and 0.2 respectively. We prepared five different split of train, validation and test dataset and conducted 5 independent trials for each training setting. After five independent training, we applied Welch's t-test [30] to test the statistical significance of each training setting compared to baseline. To correct the p-values for multiple tests we applied [4]. We excluded settings with decrease of relative improvement over 4% from statistical testing.

### 4.3 Model architecture

As a DNN model to classifiy ECG data, we adopted a convolutional neural network. Analogous to the work of [1], [18] and [23], our model employs shortcut connections which is similar to Residual Network [20] to allow information to propagate well in very deep models. Fig. 1a shows a high-level architecture of the network used for ECG classification. The network takes a time-series of raw ECG signals as an input, and outputs a label prediction for each input ECG sequence. Our model consists of 15 ConvBlocks, shown as in Fig. 1b, and followed by a fully connected layer and a softmax activation. Each ConvBlock has 2 convolutional layers with filter length of 8 and have 64k filters, where k is 1 at start and incremented every 2 ConvBlocks. Max pooling layer in ConvBlock is applied in every 2 blocks, beginning from the first ConvBlock.

We applied both Batch Normalization [21] and Dropout [28] before convolution in each ConvBlock, and inserted a rectified linear unit, ReLU after every Batch Normalization layer. The final fully connected layer and softmax activation outputs a vector of length 3, which corresponds to the number of classes in the classification task.

The details of the training setting is as follows. We used Adam optimizer [22] to train the model with a learning rate of 0.001 for both PhysioNet17 and CPSC18 dataset. We chose 10.0 as a value for gradient clipping and a batch size of 256. We trained each model for 250 epochs for PhysioNet17 dataset and 500 epochs for CPSC18 dataset, and picked a model with the best validation score and applied it on a test set to report the final score for each setting. To mitigate class imbalance problems, we computed the proportion of Normal classes in each train set for AF and Other rhythm class, and took the inverse of the value as the weight of the class.

## 5 Results

First, to test the impact of individual augmentation operations on classification accuracy, we compared the model trained with single augmentation operation on Table 1 applied respectively with and without augmentation. Experiments were conducted using both PhysioNet17 and CPSC18 dataset, with magnitude value $M$ of $[1, 2, 5, 10, 15, 20]$ for each operation respectively. Results are shown in Fig. 2. For both datasets, we observed statistically significant relative improvements of over 1.5% with shift operation with suitable magnitude values. However, for other augmentation operations on both datasets, we could not observe statistically significant improvements compared to the baseline model. These results suggest, single augmentation operation is insufficient to obtain enough variety and amount of data.

Next, to verify the effect of increasing variety of data, we applied all augmentation operations listed on Table 1. The results for both PhysioNet17 and CPSC18 datasets are shown in Table 2. For the PhysioNet17 dataset, we observed improvements with magnitude value of $M = 1, 2, 5, 10, 15$, reaching peak score of 0.8503 at $M = 5$, which was statistically significant 3.43% improvement over baseline. For the CSPC18 dataset, results with magnitude value of $M = 1, 2, 5$ improved score over baseline, with relative improvement up to 1.81%, however the results were not statistically significant compared to baseline. These results suggest, applying more augmentation operations improves classification accuracy compared to single operation applied.

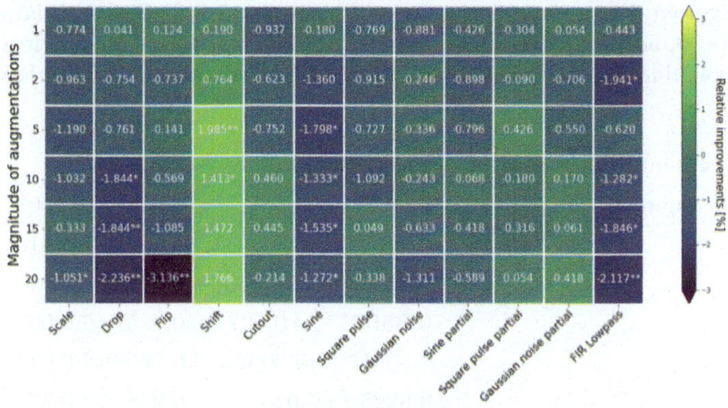


(a) PhysioNet17

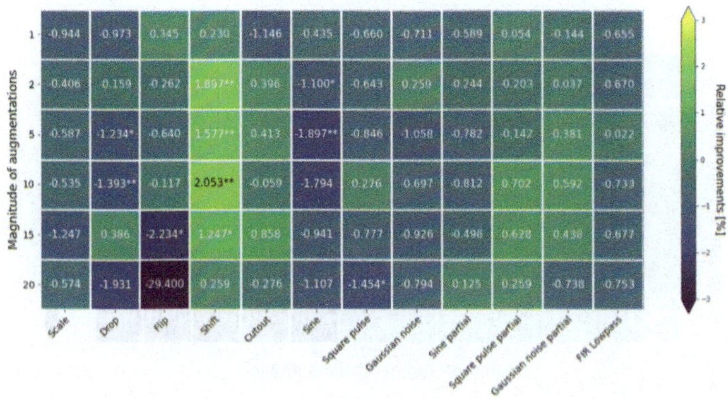


(b) CPSC18

**Fig. 2.** Relative improvement of average F1scores of 3 classes with single augmentation operation over baseline models. Fig. a and b shows results of PhysioNet17 dataset and CPSC18 respectively. ** denotes p-value below 0.01 and * denotes p-value below 0.05, p-values are corrected by [4].

Furthermore, we experimented the impact of flexibly changing the number of augmentation operations, by applying RandECG. RandECG randomly combines augmentation operations shown in Table 1, with an introduction of two additional hyperparameters: the magnitude of augmentation $M$ and the number of augmentations used at a time $N$. We conducted experiments with $M$ value of $1, 2, 5, 10, 15, 20$ and $N$ value of 1 to 12. The results for the PhysioNet17 data and CPSC18 data are shown in Fig. 3a and 3b respectively. In the experiments with PhysioNet17 data, the highest F1score was achieved when $M = 5$ and $N = 10$, a statistically significant improvement of 3.47% over the baseline model. In addition, for all models trained with $M$ value smaller than 10 or $N$ value smaller than 6, F1score was improved over the baseline model. In the experiments using the CPSC18 data set, the highest F1score was achieved when

**Table 2.** Average F1 score of 3 classes on PhysioNet17 and CPSC18 dataset with all augmentation operation applied. ∗∗ denotes p-value below 0.01 and ∗ denotes p-value below 0.05, p-values are corrected by [4]. Values inside brackets are standard deviation.

| Augmentation magnitude | PhysioNet17 | CPSC18 |
|---|---|---|
| No augmentation | 0.8221 (±0.0061) | 0.8182 (±0.0057) |
| 1 | 0.8482** (±0.0088) | 0.8242 (±0.016) |
| 2 | 0.8393** (±0.0033) | **0.8330** (±0.0086) |
| 5 | **0.8503**** (±0.0077) | 0.8257 (±0.0075) |
| 10 | 0.8361* (±0.0067) | 0.7958 (±0.014) |
| 15 | 0.8228 (±0.015) | 0.6385 (±0.22) |
| 20 | 0.6310 (±0.15) | 0.3303 (±0.049) |

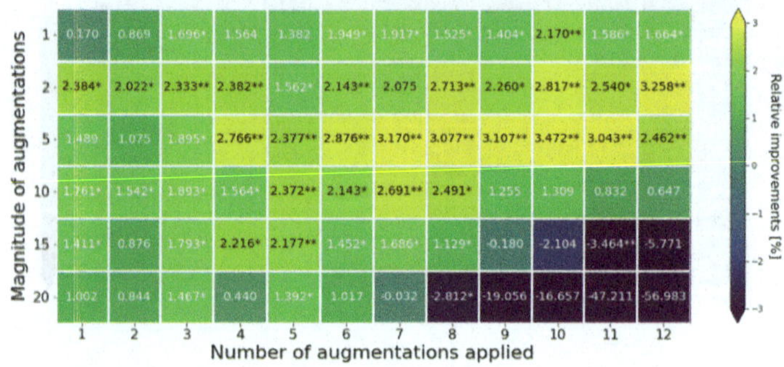


(a) PhysioNet17

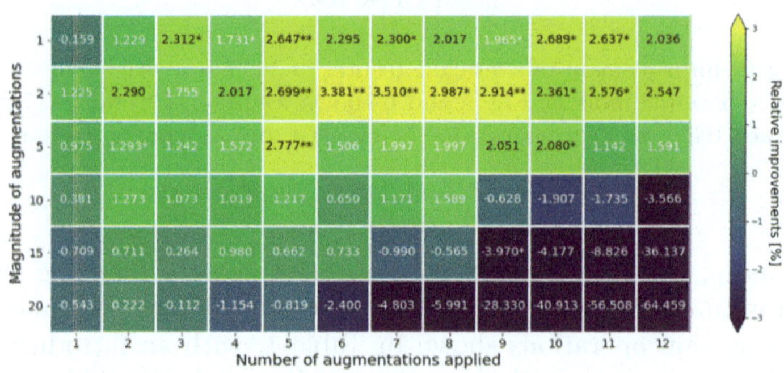


(b) CPSC18

**Fig. 3.** Relative improvement of average F1scores of 3 classes with RandECG over the baseline models. Fig. a shows results of PhysioNet17 dataset and results with CPSC18 dataset is shown in Fig. b. ∗∗ denotes p-value below 0.01 and ∗ denotes p-value below 0.05, p-values are corrected by [4].

$M = 2$ and $N = 5$, with a statistically significant improvement of 3.51% compared to the baseline. In the CPSC18 dataset, as the magnitude of augmentation increased, the F1score decreased compared to the model trained without augmentation. For both datasets, suitable $M$ value was the same to the result with all augmentation applied. These results suggest that RandECG, which combines augmentation operations randomly, can improve the classification accuracy of ECGs with suitable hyperparameters.

## 6 Discussion and conclusion

In this study, we proposed RandECG, which randomly combines augmentation operations tailored for ECG, and showed efficacy of RandECG on ECG classification tasks with DNNs. Specifically, with suitable hyperparameters selected by grid search, we observed relative improvement of classification accuracy up to 3.51% over baseline model trained without data augmentation. The model trained with RandECG achieved better performance compared to baseline model trained with single and all augmentation on Table 1 and without augmentation, on two different datasets. The optimal hyperparameters $M$ and $N$ were different for the two datasets. This may be due to differences in the ECG collection environment in the two data sets. The PhysioNet17 dataset was collected on consumer devices, whereas the CPSC18 dataset was collected in the intensive care unit. Data collected through consumer devices are likely to be noisier than in an environment controlled by healthcare professionals. Therefore, the PhysioNet17 may have been more suitable for intense augmentation operations compared to the CPSC18 dataset when performing RandECG.

In principle, the RandECG is applicable to any kind of ECG classification task, however, actual efficacy was verified only on classification of atrial fibrillation with lead I ECGs. Abnormal heart activity appears in various forms, and some transformations listed in Table 1 may disrupt classification accuracy. In such cases, the accuracy may be improved by using the AutoAugment [9], which searches for suitable combinations of augmentation operations by reinforcement learning.

The main contribution of this paper is the proposal of RandECG, flexibly combining augmentation operations. In the image domain where application of DNNs are intensively studied, data augmentations are applied not only to improve classification accuracy but also to contrastive learning [7] and semi-supervised learning [31]. The successful result with RandECG may gives opportunities to apply these frameworks to ECG data.

## References

1. Andreotti, F., Carr, O., Pimentel, M.A., Mahdi, A., De Vos, M.: Comparing feature-based classifiers and convolutional neural networks to detect arrhythmia from short segments of ecg. In: 2017 Computing in Cardiology (CinC), pp. 1–4. IEEE (2017)

2. Attia, Z.I., et al.: An artificial intelligence-enabled ecg algorithm for the identification of patients with atrial fibrillation during sinus rhythm: a retrospective analysis of outcome prediction. Lancet **394**(10201), 861–867 (2019)
3. Bahdanau, D., Cho, K., Bengio, Y.: Neural machine translation by jointly learning to align and translate. In: Bengio, Y., LeCun, Y. (eds.) 3rd International Conference on Learning Representations, ICLR 2015, San Diego, CA, USA, 7–9 May 2015, Conference Track Proceedings (2015). http://arxiv.org/abs/1409.0473
4. Benjamini, Y., Hochberg, Y.: Controlling the false discovery rate: a practical and powerful approach to multiple testing. J. Roy. Stat. Soc. Ser. B (Methodol.) **57**(1), 289–300 (1995)
5. Bowles, C., et al.: Gan augmentation: Augmenting training data using generative adversarial networks. arXiv preprint arXiv:1810.10863 (2018)
6. Cao, P., Li, X., Mao, K., Lu, F., Ning, G., Fang, L., Pan, Q.: A novel data augmentation method to enhance deep neural networks for detection of atrial fibrillation. Biomed. Signal Process. Control **56**, 101675 (2020)
7. Chen, T., Kornblith, S., Norouzi, M., Hinton, G.: A simple framework for contrastive learning of visual representations. In: International Conference on Machine Learning, pp. 1597–1607. PMLR (2020)
8. Clifford, G.D., et al.: Af classification from a short single lead ecg recording: the physionet/computing in cardiology challenge 2017. In: 2017 Computing in Cardiology (CinC), pp. 1–4. IEEE (2017)
9. Cubuk, E.D., Zoph, B., Mane, D., Vasudevan, V., Le, Q.V.: Autoaugment: learning augmentation strategies from data. In: Proceedings of the IEEE/CVF Conference on Computer Vision and Pattern Recognition, pp. 113–123 (2019)
10. Cubuk, E.D., Zoph, B., Shlens, J., Le, Q.V.: Randaugment: practical automated data augmentation with a reduced search space. In: Proceedings of the IEEE/CVF Conference on Computer Vision and Pattern Recognition Workshops, pp. 702–703 (2020)
11. Devlin, J., Chang, M., Lee, K., Toutanova, K.: BERT: pre-training of deep bidirectional transformers for language understanding. CoRR abs/1810.04805 (2018). http://arxiv.org/abs/1810.04805
12. DeVries, T., Taylor, G.W.: Improved regularization of convolutional neural networks with cutout. arXiv preprint arXiv:1708.04552 (2017)
13. Golany, T., Radinsky, K.: Pgans: personalized generative adversarial networks for ecg synthesis to improve patient-specific deep ecg classification. In: Proceedings of the AAAI Conference on Artificial Intelligence, vol. 33, pp. 557–564 (2019)
14. Golany, T., Radinsky, K., Freedman, D.: Simgans: simulator-based generative adversarial networks for ecg synthesis to improve deep ecg classification. In: International Conference on Machine Learning, pp. 3597–3606. PMLR (2020)
15. Goldberger, A.L., et al.: Physiobank, physiotoolkit, and physionet: components of a new research resource for complex physiologic signals. Circulation **101**(23), e215–e220 (2000)
16. Goodfellow, I., et al.: Generative adversarial nets. In: Advances in Neural Information Processing Systems, pp. 2672–2680 (2014)
17. Graves, A., Mohamed, A.R., Hinton, G.: Speech recognition with deep recurrent neural networks. In: 2013 IEEE International Conference on Acoustics, Speech and Signal Processing, pp. 6645–6649. IEEE (2013)
18. Hannun, A.Y., et al.: Cardiologist-level arrhythmia detection and classification in ambulatory electrocardiograms using a deep neural network. Nat. Med. **25**(1), 65 (2019)

19. Hatamian, F.N., Ravikumar, N., Vesal, S., Kemeth, F.P., Struck, M., Maier, A.: The effect of data augmentation on classification of atrial fibrillation in short single-lead ecg signals using deep neural networks. In: ICASSP 2020–2020 IEEE International Conference on Acoustics, Speech and Signal Processing (ICASSP), pp. 1264–1268. IEEE (2020)
20. He, K., Zhang, X., Ren, S., Sun, J.: Deep residual learning for image recognition. In: Proceedings of the IEEE Conference on Computer Vision and Pattern Recognition, pp. 770–778 (2016)
21. Ioffe, S., Szegedy, C.: Batch normalization: accelerating deep network training by reducing internal covariate shift. In: International Conference on Machine Learning, pp. 448–456. PMLR (2015)
22. Kingma, D.P., Ba, J.: Adam: a method for stochastic optimization. In: Bengio, Y., LeCun, Y. (eds.) 3rd International Conference on Learning Representations, ICLR 2015, San Diego, CA, USA, 7–9 May 2015, Conference Track Proceedings (2015). http://arxiv.org/abs/1412.6980
23. Li, X., Qian, B., Wei, J., Zhang, X., Chen, S., Zheng, Q.: Domain knowledge guided deep atrial fibrillation classification and its visual interpretation. In: Proceedings of the 28th ACM International Conference on Information and Knowledge Management, pp. 129–138 (2019)
24. Park, D.S., Chan, W., Zhang, Y., Chiu, C., Zoph, B., Cubuk, E.D., Le, Q.V.: Specaugment: a simple data augmentation method for automatic speech recognition. In: Kubin, G., Kacic, Z. (eds.) Interspeech 2019, 20th Annual Conference of the International Speech Communication Association, Graz, Austria, 15–19 September 2019, pp. 2613–2617. ISCA (2019). https://doi.org/10.21437/Interspeech.2019-2680
25. Porumb, M., Stranges, S., Pescapè, A., Pecchia, L.: Precision medicine and artificial intelligence: a pilot study on deep learning for hypoglycemic events detection based on ecg. Sci. Rep. **10**(1), 1–16 (2020)
26. Sandfort, V., Yan, K., Pickhardt, P.J., Summers, R.M.: Data augmentation using generative adversarial networks (cyclegan) to improve generalizability in ct segmentation tasks. Sci. Rep. **9**(1), 1–9 (2019)
27. Sennrich, R., Haddow, B., Birch, A.: Improving neural machine translation models with monolingual data. arXiv preprint arXiv:1511.06709 (2015)
28. Srivastava, N., Hinton, G., Krizhevsky, A., Sutskever, I., Salakhutdinov, R.: Dropout: a simple way to prevent neural networks from overfitting. J. Mach. Learn. Res. **15**(1), 1929–1958 (2014)
29. Szegedy, C., et al.: Going deeper with convolutions. In: Proceedings of the IEEE Conference on Computer Vision and Pattern Recognition, pp. 1–9 (2015)
30. Welch, B.L.: The generalization of student's' problem when several different population variances are involved. Biometrika **34**(1/2), 28–35 (1947)
31. Xie, Q., Luong, M.T., Hovy, E., Le, Q.V.: Self-training with noisy student improves imagenet classification. In: Proceedings of the IEEE/CVF Conference on Computer Vision and Pattern Recognition, pp. 10687–10698 (2020)

# Integration and Interaction of Trustworthy AI in a Virtual Coach – An Overview of EU-Japan Collaboration on Eldercare

Kristiina Jokinen[1(✉)], Keiko Homma[2], Yoshio Matsumoto[2], and Ken Fukuda[1]

[1] Artificial Intelligence Research Center, The National Institute of Advanced Industrial Science and Technology (AIST), Tokyo, Japan
{kristiina.jokinen,ken.fukuda}@aist.go.jp

[2] Human-Augmentation Research Center, The National Institute of Advanced Industrial Science and Technology (AIST), Kashiwa, Japan
{keiko.homma,yoshio.matsumoto}@aist.go.jp

**Abstract.** This article describes work on developing a virtual coach for assisting older adults in their everyday activities and in particular, for maintaining their well-being and healthy life. The work has been carried out in the context of the collaboration project *e-VITA EU-Japan Virtual Coach for Smart Aging*. The goal of the project is to prototype virtual coaches to sustain older adults' well-being in smart environments, to develop standards and norms for the component interoperability, and to conduct a proof-of-concept study to assess user-acceptance in real-life environments. Our focus is especially on the interaction aspects enabling the users to conduct dialogue interaction with the Virtual Coach and the integration of various system components into a federated AI platform which supports the development of the Virtual Coach.

**Keywords:** Human-robot interaction · Digital coaching · Wellbeing · Eldercare

## 1 Introduction

AI technology has already had huge impact on the human life. Applications are being developed to address challenges of the current society: demographic changes, environmental changes, and changes in work and transportation, in the ways we connect and socialize with others, etc. Technology readiness of social robotics has already proved real-world interactive applications feasible. In the context of Active and Healthy Ageing (AHA) such agents have potential to empower older adults to better manage their own health and daily activities, as well as to assist caregivers in their tasks such as early risk detection and tailored intervention. Innovative solutions are designed to improve human well-being and quality of life. However, most applications support only limited spoken language interaction and the users need to follow strict interaction protocols which can be frustrating if the interaction encounters problems in recognizing the user's speech or intention. Moreover, the applications cannot be easily customized to individual needs, or tailored according to individual interests and preferences. This reduces their usability


Y. Takama et al. (Eds.): JSAI 2021, AISC 1423, pp. 190–200, 2022.
https://doi.org/10.1007/978-3-030-96451-1_17

in daily life routines since users have their own schedules and requirements which shape the interactions and in order to provide proactive help, e.g. concerning daily reminders, it is important that the system can learn such parameters from the user. Also, integration of various devices and sensors is not fully discussed, i.e. what devices provide good user experience and what kind of sensors are useful for the task completion. Finally, data sharing and ethical issues have become an important issue in technology development of the interconnected devices, and although awareness of the issues is widely discussed, in practical application development they are still insufficiently considered.

In this article we discuss these issues in the context of the recently started project e-VITA which is a 3-year cooperation between EU and Japan (2021–2023). The main objectives of the project concern prototyping of virtual coaches to sustain older adults' well-being in smart living environments, developing norms and standards for interoperability, and conducting a proof-of-concept study to assess user-acceptance in real-life environments. For this, a modular approach will be used deploying the federated data AI platform FIWARE, and special attention is paid to standards and interoperability among the coaching components, contributing to general work on standardisation. The project will also explore big data analytics, knowledge-graphs and social-emotional computing to enable interaction that addresses relevant issues like cognition, physical activity, mood, social interaction, and spirituality.

The article is structured as follows. We first survey issues in Active and Healthy Aging, including virtual coaching and coaching projects in EU and Japan in Sect. 2. We then describe the e-VITA project in Sect. 3 focussing on interaction, integration and standardization. We discuss Trustworthy AI in Sect. 4, and finally, draw conclusions and present future work in Sect. 5.

## 2 AI for Active and Healthy Aging

In the context of current demographic situation where the increased lifespan has led to a high number of older adults, the society is faced with an expanding problem to provide daily assistance and care for the older citizens and support their wellbeing and healthy aging. The situation is also an opportunity to make full use of the advanced technology in order to foster emotional and social wellbeing through AI. Consequently, in Europe and Japan, framework projects invite research and development on new technology to support well-being and general good for the citizens. For instance, Japan presented Robot Strategy in 2015, and one of the strategies was health care and advancements in robot care indicating the growing significance of robotics and robot health care in Japan. Also, the concept of Society 5.0, which Japan introduced in 2017, envisages that humans and robots can live together in a symbiotic relation which supports human everyday activities and allows anyone to enjoy a high quality of life. The vision emphasizes that digitalization occurs in all areas of societal life, not only in industrial production, and it is to be accomplished by fostering innovation and incorporating advanced technologies in diverse industries and social activities to create new value. In human-centered society, trustworthy AI can thus be used to solve societal problems and enable new structures which will benefit all citizens. A framework for increased self-management is important for the individuals' quality of life, and it has a strong socio-economic impact on the society as a whole.

The wish to experience wellbeing is fundamentally human, and thus our approach deploys a holistic viewpoint where psychological needs of older adults are addressed by interactive technologies and assistive coaching systems. Figure 1 shows the socio-technology approach applied in e-VITA. Different stakeholders in industry, community, as well as in research and education, interact with older adults and together form a network that supports the older adults' meaningful activity, learning, and enjoyable usage of technology. An important starting point for the virtual coaching system is a framework which puts the older person in the centre of the interactions and encourages increased self-management of emotional health, social connectedness and well-being. Moreover, virtual coaches aiming to foster emotional and social wellbeing via AI also need to interact with the smart environment.

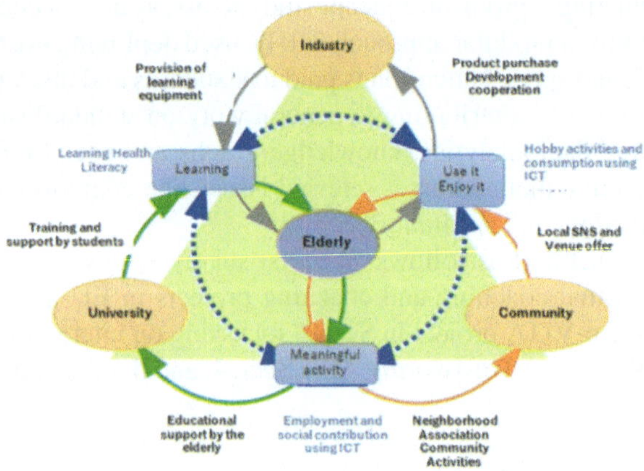


**Fig. 1.** Socio-technology approach in e-VITA. Multi-stakeholder Approach (reworked from MIC 2007).

## 2.1 Virtual Coaching

Coaching in general refers to support related to person's learning, growth, wellness, self-awareness, career management, and behavioral change. According to [6], an e-coaching system is an artificial entity that has the ability to observe a user's behaviours and use these observations as a basis for engaging in collaborative conversations with the user, in order to aid planning and promoting the user's goals through persuasive techniques. Such e-coaches have challenging qualities, with the minimal set of features consisting of *social ability* to engage in an ongoing conversation with the user, *credibility* to be perceived as having expertise and being trustworthy, *awareness of context* to stimulate ideas and action, *learning ability* to ask pertinent questions, give feedback, and offer advice that is tailored to the individual user. Moreover, it should also be able to *interface with data streams* to gather different types of data (user input, measurements of physical activities, mood self-reports, sleeping patterns, etc.), *be proactive* to initiate interactions with the user to stimulate actions, provide reminders, issue warnings, etc. Most importantly, it should have a *model of behaviour change* to support coaching activities, i.e., not only act as an instrument but behave as a coach. Consequently, it should be able to develop *strategies* for behaviour change to guide the user in a process, be *capable of planning* future-directed intentions.

### 2.2 Virtual Coaches in Japan and EU

As part of the Cabinet Office's "Dynamic Engagement of All Citizens" plan, several recent projects have studied issues how to maintain and improve the quality of life for older people through support for their independent life.

For instance, the NEDO2.0 Future AI and Robot Technology Research and Development Project (2015–2020) focused both on basic research and social implementation, the former aiming to develop core technology for next generation artificial intelligence and robotics, and the latter aiming to solve concrete problems by the next generation AI technology. The METI (Ministry of Economy, Trade and Industry)/AMED (Japan Agency for Medical Research and Development) Project to Promote the Development and Introduction of Robotic Devices for Nursing Care (2013–2017) aimed at developing nursing care devices and evaluation methods for safety and effects of the developed robotic devices, determining guidelines for the development and evaluation, and establishing standards for the devices.

Several European projects also have focus on virtual coaching. For instance, the H2020 project EMPHATIC implements coaching through a virtual conversation agent. The system provides personalised advice and guidance with a direct impact on empowering older users into a wide use of advanced ICT thus improving their quality of life and level of their independent living status. The dialogues follow a sequential structure based on the GROW model (Goal, Reality, Options and Will of action plan) which was proposed by health coaching professionals. Another project, NESTORE, developed a virtual coach to sustain motivation to remain healthy, while the projects CAPTAIN and SAAM focused on developing personalized virtual coaches to assist older adults living independently.

In the context of EU-Japan collaboration, the project CARESSES (Culture Aware Robots and Environmental Sensor Systems for Elderly Support) is a prior project funded by European Union and the Ministry of Internal Affairs and Communications of Japan. It focused on cultural competence of robots, i.e. making robot technology more acceptable and sensitive to the user's needs, customs and lifestyle, by addressing the question of how the robot can recognize various cultural factors and take different values and beliefs into consideration in its behaviour. Culture-aware assistive technologies were developed in smart environments where the user can manage interconnected devices and sensor equipment, e.g. lighting, temperature etc. by interaction.

## 3 E-VITA Virtual Coach

### 3.1 E-VITA Project

The project *e-VITA Virtual Coach for Smart Aging* is an EU-Japan collaboration project, and its main objectives concern prototyping of a virtual coaching system in smart living environments, to support older adult's everyday activities, independent living, and well-being. The project also aims at developing norms and standards for interoperability and conducting proof-of-concept studies to assess user-acceptance in real-life environments. The project has 22 participants from Europe (Germany, France, Italy, Belgium) and Japan, and is funded by European Union and the Ministry of Internal Affairs and Communications of Japan. The interdisciplinary consortium deploys socio-technology approach of participatory design, cf. Figure 1, and builds around everyday practices via real-life living labs. The expected impacts of the project are:

1. to develop and evaluate a new ICT-based coaching system for smart living and for active ageing,
2. to motivate older adults in improving their lifestyle by promoting short-term gains with mid-term aims in order to achieve long-term behavioural changes and improved quality of life and well-being,
3. to provide a modular system for the integration of risk assessment and intervention services for use in existing platforms that can be used by different suppliers,
4. to provide collaboration between Europe and Japan to establish smart living environments for ageing well.

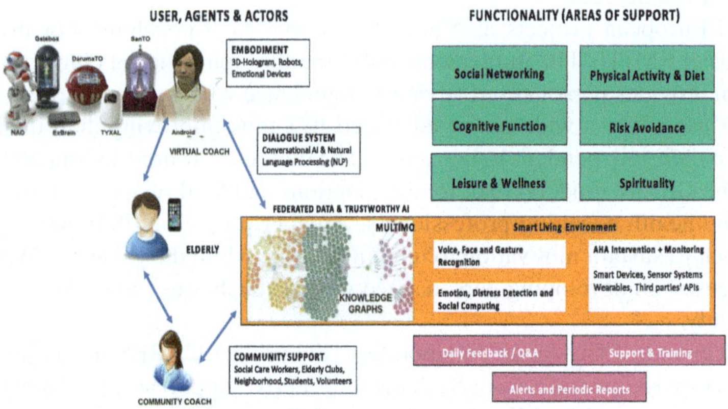


**Fig. 2.** e-VITA coaching system with interface agents.

## 3.2 System Overview and Integration

The functionalities of the coaching system are shown in Fig. 2. They are expected to provide a wide range of support from cognitive functions to leisure, wellbeing, and spirituality in smart environments. Coaching activities are expected to include human interaction with the coaching agents but also contribute to community support by engaging volunteers, social care workers and the neighbourhood in the building of a coaching environment that enables symbiotic human and technology co-existence.

The virtual coach can have different embodiments. The project plans to use a variety of interactive devices and explores e.g. the humanoid Nao, a common platform for human-robot interaction studies from Softbank, the Gatebox with a 3D holographic structure to display the virtual coach, two spiritually oriented social robots Daruma and Santo, as well as the Android [10] which has human appearance. Our focus is on human-robot interaction and we plan to use humanoid robots, in particular, Android and Softbank's Nao robot, in the data collection, development, and evaluation studies.

The core technologies to realize virtual coach and to support the above functionalities consist of a federated data AI platform, sensors, and trustworthy AI technology, especially knowledge graphs [3] and dialogue modelling [5]. Knowledge graphs provide storage for the coaching knowledge and a dialogue system which enables interaction between the virtual coach and human users. Smart living environment provides sensor-based information which is important for determining appropriate system responses in different environmental situations.

The technical platform consists of device related components (gateways, IoT Agents and Platforms), interoperability standards and interfaces (device interfaces and data management APIs), data related components as well as data analytics and AI components. A more detailed view of the system components is given in Fig. 3.

e-VITA coaching system is required to integrate various components such as devices and data analyzing and processing functionalities. Also, the system should provide interoperability and scalability for future use. To meet these requirements, FIWARE [2] considered as a basement of the e-VITA platform. FIWARE is an open-source framework for data distribution developed in EU.

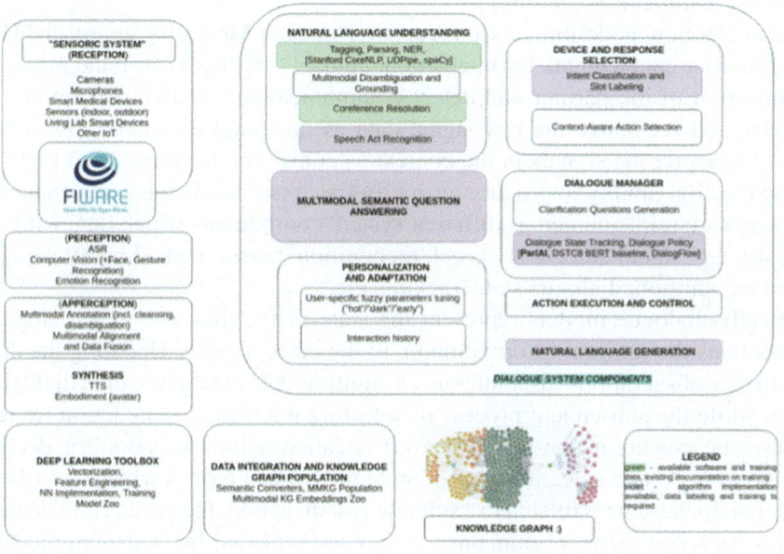


**Fig. 3.** The overall system component architecture with a focus on interaction management.

As depicted in Fig. 3, the development of e-VITA coach consists of several tasks, set to be achieved in cooperation with the partners. The tasks include e.g. natural language understanding and generation, device and response selection, core dialogue management, personalisation and adaptation, as well as knowledge-base actions for graph modelling, deep-learning toolbox for data analytics as well as multimodal emotional and social computing for the user's situation assessment.

### 3.3 Interaction

The e-VITA coaching system's appearance is related to the different embodiments of the interface agents planned to be used in the project: Nao, Gatebox, Android, and Daruma. Appearance is expected to have impact on the user's views of the interactions in that certain type of dialogue activities are assumed to be more natural with some robots, like talking about spiritual aspects with Daruma, doing exercises with Nao, or conducting engaging interactions with Gatebox or Android. It is possible that the interface agents will not use all of the conversational capabilities available, but the most suitable activities for a certain activity can be selected based on the data collection and users' views about natural dialogues and suitable appearance in the living lab experiments, i.e. the robot's appearance may afford certain type of dialogue interactions most natural in the given context. However, the system's dialogue capability will be the same across the devices, thanks to the general platform that integrates the dialogue manager with the other system components and devices. In other words, the appropriate appearance of the coach is assumed to be partly due to the user's personal preferences and partly due to the usability of the coach as experienced through the interaction with a particular interface agent.

*Constructive Dialogue Modelling.*

We base interaction modelling on Constructive Dialogue Modelling approach [4] which emphasizes four enablements for interaction: contact (ability to sense the presence and communication of the partner and detect communicative signals), perception (ability to perceive and recognize the raw signal data as meaningful signals), understanding (ability to interpret the signals in the context as conveying intentions and topical entities), and reaction (ability to produce an appropriate response in the given context). The enablements are implemented as different system components which deal with the relevant tasks, e.g. signal detection, speech recognition, parser, named entity recognition, intention recognition, dialogue state tracking, etc.

Typically dialogue models represent the state of the dialogue in a dialogue state which contains the relevant information up to the current point. Updating the dialogue state with new observations (user utterances, multimodal signals) is called dialogue state tracking, while the subsequent process of selecting the next system action (to respond to the user or execute a non-verbal action) is called action selection, or decision of what to do next. Action selection can be implemented using rules which describe how a particular dialogue state is related to a suitable system action. However, rules do not scale up well to larger or different domains, so they are replaced by statistical models, and recently by neural models. Although neural networks have been successful in developing end-to-end conversational AI systems, leading to enhanced performance across a wide range of dialogue processing tasks, there are still issues related to coherent task-based interactions as opposed to big-data enabled chatting, as well as general evaluation and ethical data processing requirements.

*Rasa Open-source Conversational Framework.*
The dialogue manager is implemented on the basis of the Rasa Open Source Framework which offers technology to develop conversational agents and supports both rule and machine learning approach. The Rasa architecture consists of the NLU Pipeline and Dialogue Policy components, with the DIET (Dual Intent and Entity Transformer) forming a state-of-the-art multitask transformer architecture for NLU to recognize the user intent and extract relevant entities, while transformer-based machine learning is incorporated in the Rasa dialogue policies as Transformer Embedding Dialogue Policy (TED Policy), to predict the next action using the attention mechanism (to pay attention to and selectively ignore dialogue turns when making predictions of the next action). Rasa custom actions can be used to access the outside world (e.g. database access, robot commands), and they are executed through Action Manager. Rasa operates on text input, so spoken interaction requires integration of separate speech technology components: ASR to recognize speech signals and produce transcription of the speech, and TTS to get the text message spoken to the user.
*Co-creation of a dialogue system and standardisation.*
The socio-technological view and co-creation of the coach application is realised by early experimentation of the dialogue model with older users. This accords with the principles of Conversation-Driven Development (CDD) which is the overarching best practice approach for chatbot development in RASA. Real users can provide language and behaviour insights to improve the system early on.

A crucial part of dialogue modelling is related to the number and type of user intents for a given dialogue. While most dialogue systems usually resort to a suitable set designed for the task at hand with "intents" and "entities" specific for the application, there are efforts on standardisation of dialogue acts. For instance, [1] aims at a comprehensive view of dialogue acts, which is continued in the ISO (International Organization for Standardization) work.

We aim to address dialogue intent design as part of the general standardisation and interoperability issues, and thus consider the intent design from the point of view of generalising or reusing the intents in other domains, too. A starting point is the first implementations of the dialogue intents and entities for the given use cases and inviting the users to use the system from early on. The purpose is to learn how the users talk and interact with the system so as to collect data for further development of the dialogue manager's natural interaction capability and enable early co-creation of the coach with the real user interaction. The guiding principle is to model dialogue interactions in a flexible manner which allows generalisation and reusability of the models in a variety of use cases, input configurations and user needs.
*Interaction with the environment.*
Some of the environmental sensors that the e-VITA project plans to use concern measuring room temperature and detecting movement in the room, as well as infra-red distance detector on the robot agent. Room temperature influences the reminders that the coach provides to the user, while the movement detectors will especially focus on the initialisation of the dialogues and take into account the Contact enablement of CDM [4]. Contact is the first enablement of dialogues and refers to the partners being aware of each other's presence and in the proximity for being able to start interaction. For instance, the user

and the coach must be in the same space and in such close contact that spoken dialogue can start.

Other environmental sensors are also planned to be used in the project, for instance a domotic sensor network of motion tracker, temperature, and light sensors. The influence of such sensor networks on dialogue management is expected to relate e.g., to the user's comfort and well-being, and will be explored through specific experiments along the lines of [8] using Heating, Ventilation and Air Conditioning (HVAC) control loops in particular interactive situations.

### 3.4 Standardisation

The project will also work on standards, norms and interoperability of federated data and devices in Smart Living Systems. This includes stake holder engagement and sustainability of smart living solutions for Active Healthy Ageing (AHA) in Europe and Japan. In a previous project, consortium members AIST and JQA (Japan Quality Assurance Organization) were members of the consortium which developed safety testing technologies for service robots and built a "Robot Safety Center" where all test equipment required for service robots was installed. As a result of this project, the ISO adopted Japan's proposal and issued ISO 13482 for service robots as a new global safety standard in 2014. The adoption will enable manufacturers of such robots to gain certification in order to meet the global safety standards. It is also expected that manufacturers will be able to gain the trust of customers with a proof of certification, having met internationally recognized safety standards.

## 4 Trustworthy AI

The nature of the relationship between smart living technology and older adults has raised several opportunities, but also some concerns and challenges related to privacy and ethical issues. Many smart living functionalities require storage of data as well as the comparison and exchange of data with other systems, e.g. electronic health records or Q/A services, or they may repurpose the stored cloud data to improve their intelligence. With new advances in trustworthy AI and federated data structures and edge computing (e.g. GAIA-X), the sharing of personal data between virtual coaching systems and components of smart environments has already begun and will become more common and pervasive. It is envisaged that virtual coaches as part of such an ecosystem will go beyond their traditional role as stand-alone entities facilitating specified parts of care, and instead become nodes of complex information sharing networks between Europe and Japan.

Virtual coaches will increasingly interact with smart devices and become part of the pervasive information environment. Such decentralized and spatially distributed components present challenges regarding the use of information and in particular, data security. As humans and virtual coaches act together in a team in the living community, they utilize their respective capabilities in cooperation in order to achieve joint tasks (coactivity). In such teams, it is important to create mutual trust between all the actors. A precondition for trust developing is that the human operators can maintain mental models of their virtual machine counterparts, while simultaneously, the coaching system needs to communicate its internal state and explain its actions and decisions in an accurate manner that supports the user's understanding of the factual data. Development of trust within such coactive teams can be slow and require both technological reliability and human patience.

The e-VITA project will pay special attention to the issues on how to build reliable applications and software. We will focus on issues in data sharing, especially between countries and continents, privacy awareness, and well as control and trust issues in Federated Data in Smart Living Systems in Europe and Japan. Finally, we will discuss and aim to pave way on the important and many-sided aspects of ethical issues when designing and developing applications for older adults.

## 5 Discussion and Future Plans

Despite the expected opportunities and advanced technology, there are many challenges in the development interactive, natural and ethically trustworthy virtual coaches, and still a lack of sustainable use of such smart living systems. We believe, on the basis of our experiences, that these challenges can be solved by socio-technology oriented approaches with co-creation methods for application development, i.e. bringing together the older adults with the technology developers, and combining the technology use with community activities, thus supporting face to face interaction in groups by technology measures, as well as data collection and content creation by first-hand user experience.

Following all these considerations, an important ambition of e-VITA is to establish new socio-technological approaches in Europe and Japan and to evaluate them in real-life practices of community-dwelling older adults. It is also expected that "uncanny valley" [7] will be investigated in the project, in particular how it works using variety types of conversation robots and agents including androids. Older adults with different cultural backgrounds in Europe and Japan are planned to interact with them in the experiments.

More information about the e-VITA project, news, events, and results can be found on the website: https://www.e-vita.coach/.

**Acknowledgements.** We thank our project partners in Japan and EU for discussions and collaboration. This work was supported by the Strategic Information and Communications R&D Promotion Programme (SCOPE) of Ministry of Internal Affairs and Communications, Grant no. JPJ000595. The European consortium received funding from the European Union H2020 Programme under grant agreement n° 101016453.

## References

1. Bunt, H., et al.: The ISO standard for dialogue act annotation. In: Proceedings of the 12th Language Resources and Evaluation Conference, pp. 549–558 (2020)
2. Cirillo, F., Solmaz, G., Berz, E.L., Bauer, M., Cheng, B., Kovacs, E.: A standard-based open source IoT platform: FIWARE. IEEE Internet Things Mag. **2**(3), 12–18 (2019)
3. Ehrlinger, L., Wöß, V.: Towards a definition of knowledge graphs. SEMANTiCS (Posters, Demos, Success) **48**(1–4), 2 (2016)

4. Jokinen, K.: Constructive Dialogue Modelling–Speech Interaction and Rational Agents. Wiley, Chichester, UK (2009)
5. Jokinen, K., McTear, M.: Spoken Dialogue Systems. Morgan and Claypool (2009)
6. Kamphorst, B.A.: E-coaching systems: what they are, and what they aren't. Pers. Ubiquit. Comput. **21**(4), 625–632 (2017). https://doi.org/10.1007/s00779-017-1020-6
7. Mori, M.: The uncanny valley. Translated by MacDorman, K.F., Kageki, N. IEEE Robot. Autom. **19**(2), 98–100 (2012)
8. Morresi, N., et al.: Sensing physiological and environmental quantities to measure human thermal comfort through Machine Learning techniques. IEEE Sens. J. **21**(10), 12322–12337 (2021)
9. Wang, Q., Mao, Z., Wang, B., Guo, L.: Knowledge graph embedding: a survey of approaches and applications. IEEE Trans. Knowl. Data Eng. **29**(12), 2724–2743 (2017)
10. Yoshikawa, M., Matsumoto, Y., Sumitani, M, Ishiguro, H.: Development of an android robot for psychological support in medical and welfare fields. In: Proceedings of the IEEE International Conference on Robotics and Biomimetics (ROBIO2011), pp. 2378–2383 (2011)

## Projects

11. CAPTAIN. https://www.captain-eu.org/
12. CARESSES. http://caressesrobot.org/en/project/
13. EMPHATIC. http://www.empathic-project.eu/
14. e-VITA. http://www.evita-coach/
15. NESTORE. https://nestore-coach.eu/home
16. SAAM. https://saam2020.eu/
17. Robotic Devices for Nursing Care: http://robotcare.jp/?lang=en
18. Project to Promote the Development and Introduction of Robotic Devices for Nursing Care: https://www.amed.go.jp/en/program/list/02/01/005.html

## Other links

19. GAIA-X. https://www.data-infrastructure.eu/GAIAX/Navigation/EN/Home/home.html
20. GATEBOX. https://www.gatebox.ai/
21. NAO Robot. https://www.softbankrobotics.com/emea/en/nao
22. RASA. https://rasa.com/
23. DIET. https://rasa.com/blog/introducing-dual-intent-and-entity-transformer-diet-state-of-the-art-performance-on-a-lightweight-architecture/
24. TED. https://rasa.com/blog/unpacking-the-ted-policy-in-rasa-open-source/
25. Society 5.0. https://www.japanindustrynews.com/2019/04/connected-industries-japans-approach-to-industry-4-0-on-the-way-to-society-5-0/

# Anti-Drug Buddy: A Chatbot for Advocating the Awareness of Anti-Drug

Lieu-Hen Chen[1], Eric Hsiaokuang Wu[2(✉)], Chih-Chieh Yang[2], Ting-Yu Gao[2], and Li-Hsien Tang[2]

[1] National Chi Nan University, Puli Township, Taiwan
[2] National Central University, Taoyuan, Taiwan
hsiao@csie.ncu.edu.tw

**Abstract.** In recent years, the spread and abuse of drugs have always been the focus of social attention because it can cause serious problems in our society. Although there have been many efforts in the domain of anti-drug, the number of guilty persons in drug cases is still rising. Therefore, there is a need to find a way to promote anti-drug awareness. At present, chatbots on the market cannot be completely applied to the anti-drug domain, so we design a dialog architecture to improve the anti-drug-related QA and to answer more detailed knowledge about drugs. Then, we propose Anti-Drug Buddy that utilizes a chatbot to advocate general knowledge about drugs and deepen anti-drug awareness for people. Currently, our Anti-Drug Buddy is already serving on LINE. For an anti-drug domain, our chatbot can answer users: Information about drugs, How to deal with drug problems, Pushing the latest news and information about drugs.

**Keywords:** Anti-drug · Chatbot · Natural languages · Knowledge based systems

## 1 Introduction

The spread and abuse of drugs have always been the focus of social attention. The spread of drugs can cause serious problems in our society and endanger the lives, property, and health of the public. The United Nations Office on Drugs and Crime, UNODC 2013 World Drug Report, shows that between 3.6% and 6.9% (167 million to 315 million people) of the world's 15–64 year-old population have drug abuse problems [1]. Because of this negative impact, how to help people to recognize the harm caused by drug abuse has become an indispensable problem. Therefore, there is a need to find a way to effectively disseminating anti-drug materials to help people enhance awareness of the dangers of drugs. To solve this problem, many related researchers have tried to study how to disseminate anti-drug materials to help people realize the harm caused by drug


Y. Takama et al. (Eds.): JSAI 2021, AISC 1423, pp. 201–212, 2022.
https://doi.org/10.1007/978-3-030-96451-1_18

abuse, especially media broadcasts and educational programs, and use some drug-related educational videos to provide anti-drug information [2]. There is also a game-based anti-drug system developed to help non-addicted users achieve drug harm and prevent addicted users from taking drugs again [3,4].

Over the years, the government has carried out a lot of related work in the domain of anti-drugs. In the face of the current situation and predicament of drugs, the Executive Yuan put forward the "new generation anti-drug strategy" on May 11, 106, one of which is to integrate the local anti-drug center. The agency was changed from the Ministry of Justice to the Department of Health and Welfare, integrating the medical treatment and counseling functions of the local poison defense center. According to the statistics on the number of drug offenders provided by the Ministry of Justice in 2018 [5], although there have been many efforts in the domain of anti-drug, the number of guilty persons in drug cases is still rising. Moreover, the existing anti-drug information is too scattered to recognize the dangers of drugs for people. Therefore, there is a need to find a way to promote anti-drug awareness.

On the other hand, artificial intelligence (AI) is a hot topic in the past few years. More applications can be achieved with an artificial intelligence computer. Especially in the field of chatbots and personal assistants, AI can significantly improve the early days of Eliza, Alice. The regular chatbot, Apple Siri, Amazon's Alexa, Microsoft XiaoIce, and Cortana are also examples of artificial intelligence dialogue systems and personal assistants [6]. Google Duplex, a new technology for conducting natural conversations to carry out "real world" tasks over the phone, such as booking a restaurant [7]. Currently, Chatbot can be deployed on social media messaging platforms (Facebook messenger, Telegram, Line, etc...). These social media platforms can easily deploy their own chatbots simply by calling the API they provide. These kinds of chatbots are easy to use and very convenient for the user because they are based on conversation and messaging app that we are already familiar with.

Anti-drug-related government agencies also provide some anti-drug official social media accounts to publicize the anti-drug knowledge and solve some questions and answers on drugs for people. However, these accounts are still manual services, lacking immediacy. There are several advantages to using a chatbot instead of manual customer service: chatbots can work around the clock, effectively reducing labor costs. Therefore we utilize the chatbot to advocate general knowledge about drugs and deepening anti-drug awareness for people. [8,9], and [10] provide an overview of the current chatbot building approaches. [8] describes two of the main chatbot building architectures: rule-based approaches and data-driven approaches. Rule-based systems were exploited earlier and they rely on pattern-action rules. [11,12] and [13] are rule-based chatbots. [11] uses the open-source Artificial Intelligence Markup Language (AIML) program-o to design templates for diabetes questions and answers. [12] is a DBpedia chatbot which is a knowledge-graph-driven chatbot based on Wikipedia. In general, a

Knowledge Graph is a bunch of interrelated information, usually limited to a specific business domain, and managed as a graph. The interrelations provide new insights into the business domain. A Knowledge Graph can:

1. Describes real-world entities and their interrelations.
2. Defines possible classes and relations of entities in a schema.
3. Allows for potentially interrelating arbitrary entities with each other.
4. Covers various topical domains.

The knowledge Graph is a form of massive knowledge representation, which contains various entities, concepts, and various semantic relationships. It can be considered as a way to store a database. The knowledge Graph is composed of pieces of knowledge, each knowledge can be represented as an SPO/RDF triplet (entity - attribute - attribute value). In Knowledge-Base Question Answering (KB-QA), [14] proposed a question-answer system KBQA based on the Knowledge base. QA over knowledge base now becomes an important and feasible task, so we can utilize knowledge Graph to improve question answering and in a chatbot. Briefly, we can first extract the name of the entity that is asked in the question based on NER (Named Entity Recognition), and then get the triples associated with the entity from the knowledge graph. Finally, find the relevant attribute in the triples that are asked in the question, and respond to the attribute value as the answer.

On the other side, the data-driven approaches rely on large conversations dataset. Data-driven chatbots are the newest approaches and most used ones. When rule-based approaches use handwriting rules to respond to answers, data-driven approaches use existing conversations in the real world. In order to use the existing conversational data, we can use whether retrieval-based model or generative-based model. [15] and [16] are the retrieval-based chatbots, these models are the matching process between the input query and question of the QA-pair dataset. All the response answers are maintained and stored in the database. [17,18] are the generative-based chatbot. Unlike the retrieval-based model, the generative-based model does not store predefined responses in the database. It creates a response by the model itself so it can use in applications such as chitchat. The most used machine learning model in the generative-based model is a sequence-to-sequence model. For instance, [18] builds a chitchat bot with the conversation training datasets from twitters.

At present, the chatbot system on the market is mostly based on the retrieval model because the cost of grammatical errors in the product system is very expensive and may lead to user loss. Due to these facts, we adopt the retrieval-based model in our system. The chatbot mentioned above can not be completely applied to the anti-drug domain, so we propose other architectures to improve anti-drug QA and respond to more detailed knowledge about anti-drug. In this specific domain of anti-drug, our chatbot can answer user:

1. Information about drugs
2. How to deal with drug problems
3. Pushing the latest news and information about anti-drugs.

## 2 System Design

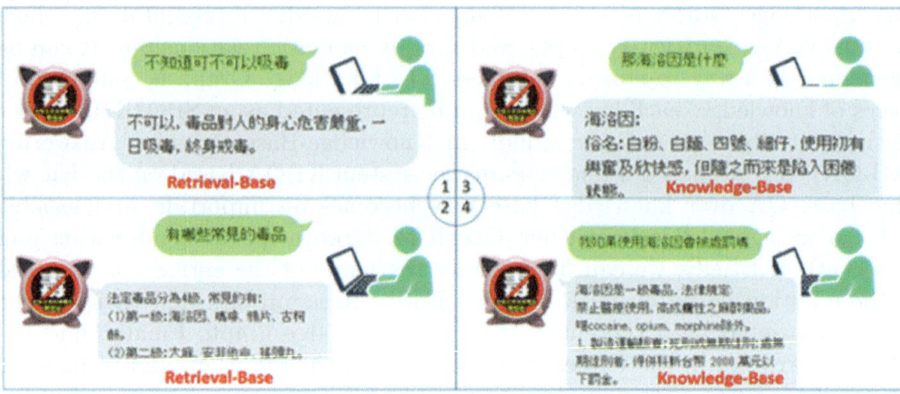


**Fig. 1.** Scenario of Anti-Drug Buddy

Figure 1 shows how Anti-Drug Buddy can help users. According to different types of problems, we need different modules to reply. The design is described as follows: We use a web server to handle the social messaging platform's API requirements and overlay balancing. After receiving the message, the web server will forward it to our chatbot system via HTTP. In the chatbot server, for user's query, we have three response modules: rule-based module, knowledge-based module, and retrieval-based module. When the chatbot server obtains an answer from the response module, sends it back to the web server and social media platform. Meanwhile, the chatbot server will save the dialogue (user's query and chatbot's response) into the database that can then be used as a corpus to improve the response.

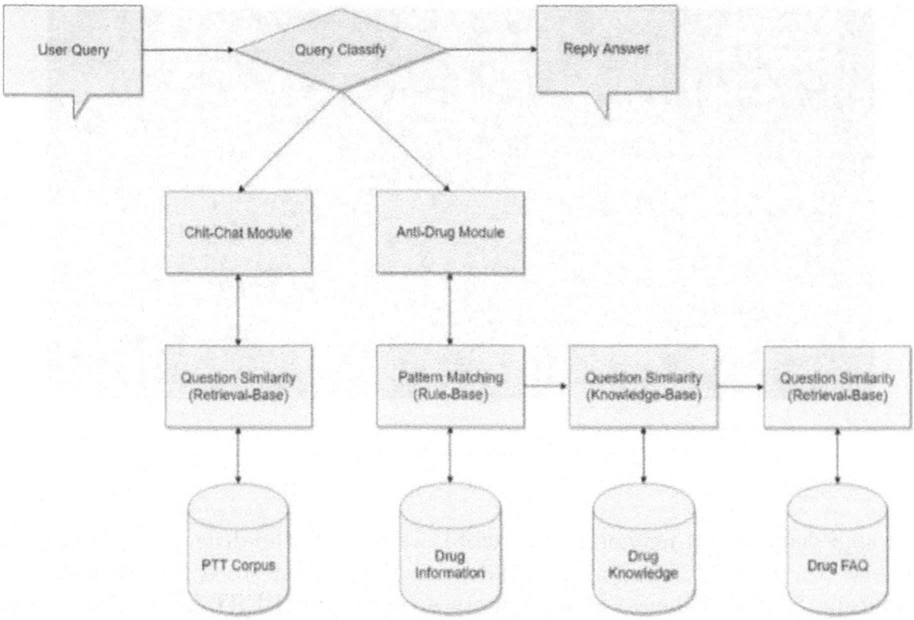


**Fig. 2.** System workflow

Figure 2 shows our proposed system workflow. The workflow is described as follows: After receiving the user's query, our system will first classify the query as if it is in the anti-drug domain. If it is in the anti-drug domain, we will use three response modules in sequence to get the answer.

In rule-based module, it is designed to answer questions based on prefixed rules. During the dialogue, the bot follows specific rules to chat with the user. We use this module to push the latest news and information about drugs that are collected as Anti-Drug information dataset, and then check the query whether it is predefined in the script (if-else/keyword) and get the response from this rule. For example, when the query is "drug news", responds the latest news about anti-drug crawled from the government or news website.

In knowledge-based module, similar to [12] and [14], we will first extract the entity from the query by keyword or Named-entity recognition (NER). Second, find all the data related to the entity from the drug knowledge graph built with Neo4j, and then retrieve the most similar attributes by keyword or vector similarity in the query related data of the entity. Finally, take the attribute value of the data as the response answer. Figure 3 shows our knowledge-based module pseudocode and related information about one of the drugs, PMMA, in our drug knowledge graph.

In retrieval-based module, we use the user's query as input and retrieve the most similar question from the drug QA-pairs dataset, then return the answer of QA-pair. First, we need to generate distributed representation from text inputs

```
# Knowledge-Base Module
drug_triplet = (...) # assume we have a drug knowledge_graph
drug_entities = [...] # assume we have a drug entity array
while True:
    if query: # get user request
        # Extract the entity from the query
        entity = get_query_entity(query, drug_entities)
        # Find all the data related to the entity from the drug knowledge graph
        entity_triplet = get_triplet(entity, drug_triplet)
        # Retrieve similar attributes in the query related data of the entity
        # use BERT encode to get sentence vector
        query_vec = BERT.encode(query)
        score = []
        for i in entity_triplet:
            attribute_vec = BERT.encode(i['attribute'])
            # Calculate similarity between query and entity attribute
            score.append(similar(query_vec, attribute_vec))
        # Find the most similar vector of question's index, and return answer
        index = Find_max_score(score)
        # Take the attribute value of the data as the reply answer
        return entity_triplet[index].value
```

**Fig. 3.** Knowledge-based module

through deep neural networks. Therefore, we convert the drug QA-pair dataset to a vector matrix, mapping a variable-length sentence to a fixed-length vector using BERT. When a user requests a query, use BERT to get this query vector, and then the degree of matching is computed as the Euclidean distance or cosine similarity between two vectors (for each question in vector matrix). Finally, we will respond the answer of QA-pair after retrieving the most similar sentence vector between two vectors. Figure 4 shows our retrieval-based module pseudocode.

```
# Retrieval-Base Module
# Convert each question to sentence vector
questions = [] # QA_pair's question array
answers = [] # QA_pair's answer array
questions_vecs = [] # QA_pair's question vector matrix
file = open("QA_pair.json")
for v in file:
    questions.append(v['question'])
    answers.append(v['answer'])
questions_vector = BERT.encode(questions)
# Retrieve the most similar of sentence vector between the user's input
# and QA dataset questions
while True:
    if query: # get user request
        query_vec = BERT.encode(query)
        score = []
        for i in questions_vector:
            # Calculate similarity between query and entity attribute
            score.append(similar(query_vec, i))
        # Find the most similar vector of question's index, and return answer
        index = Find_max_score(score)
        # Take the QA_pair's answer as the reply response
        return answers[index]
```

**Fig. 4.** Retrieval-based module

In the anti-drug domain, in order to integrate anti-drug related information, we collect drug introductions and drug-related QA information as drug dataset from Anti-drug center websites, Ministry of Justice, exam papers for students, and so on. In the drug QA-pair dataset, we have crawl more than 200 pairs of questions and answers.

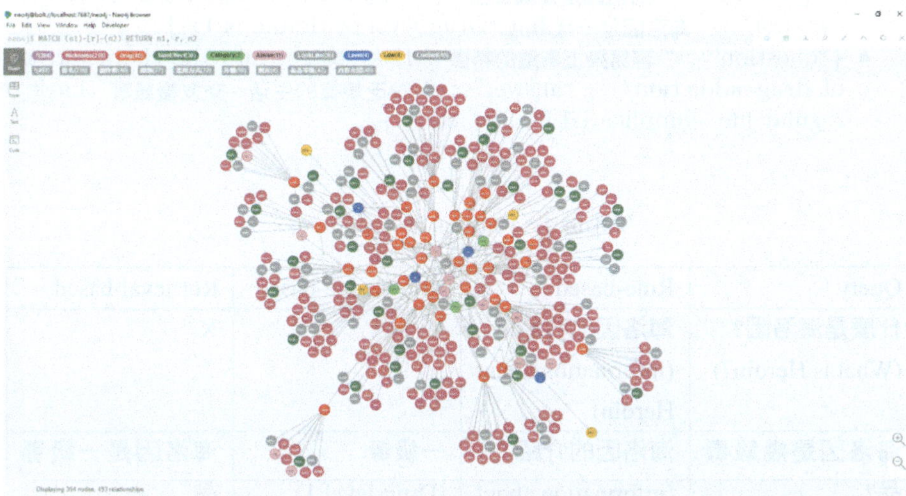

**Fig. 5.** An overview of drug knowledge graph built with Neo4j

In drug Knowledge Graph, our basic dataset is generated manually by professionals, we define 10 drug-related attributes as a necessary relationship. Figure 5 illustrates the overview of the drug knowledge graph. Orange nodes represent drugs and the others are related attributes (e.g. nickname, category, side effect, etc.). Then we collect the triplet of drugs from other places as supplementary knowledge. First, we use Fudan University's CN-DBpedia [19], which is an open-source Knowledge Graph dataset, and then we extract from other semi-structured data such as Wiki and Baidu. Our drug Knowledge Graph has about 30 drug entities with more than 600 knowledge triplets.

In Anti-Drug information, we crawl related news, articles, and the latest information from Anti-drug center websites. We collect more than 1,000 articles. However, not in the anti-drug domain, we build the Chitchat QR-pair corpus that was collected by crawler from PTT Gossiping and C_Chat. We have more than 500,000 pairs of queries and responses.

To analyze the improvement of each module in the query, we design the following scenarios, and show the result in Table 1:

1. Rule-based
   - If query contains " 海洛因 (Heroin)", return " 海洛因的介紹 (information of Heroin)"

- If query contains " 毒家新聞 (drug news)", return " 毒品的新聞 (the latest drug news)"

2. Knowledge-based
   - [海洛因 (Heroin), 毒品等級 (drug-level), 一級毒品 (drug-level 1)]
   - [海洛因 (Heroin), 俗稱 (alias), 白粉 (white powder)]
3. Retrieval-based
   - {"question" : " 海洛因是幾級毒品 ?(What is the drug-level of Heroin?)", "answer" : " 海洛因是一級毒品 (Drug-level of Heroin is 1)"}
   - {"question" : " 容易染上毒癮的特徵 ?(Easy to get into the characteristics of drug addiction?)", "answer" : " 缺乏規律的生活、交友複雜等 (Lack of regular life, complicated friends, etc.)"}

**Table 1.** Analysis of 3 modules

| Query | Rule-based | Knowledge-based | Retrieval-based |
|---|---|---|---|
| 什麼是海洛因? (What is Heroin?) | 海洛因的介紹 (information about Heroin) | X | X |
| 海洛因是幾級毒品? (What is the drug-level of Heroin?) | 海洛因的介紹 (information about Heroin) | 一級毒 (Drug-level 1) | 海洛因是一級毒品 (Drug-level of Heroin is 1) |
| 海洛因的其他名稱? (What are the other aliases for heroin?) | 海洛因的介紹 (information of Heroin) | 白粉 (Dope) | X |
| 毒家新聞 (Drug news) | 毒品的新聞 (Drug news) | X | X |
| 哪種人容易吸毒? (Who is easy to take drugs) | X | X | 缺乏規律的生活、交友複雜等 (Lack of regular life, complicated friends, etc.) |

According to the situation set in the previous, in the rule-based module, whenever there is a query containing Heroin, it will reply to the introduction of Heroin, but the answer is incorrect in the query of No. 2 and No. 3. In the knowledge-based module, it can answer more detailed knowledge when there is a related knowledge triplet. In the retrieval-based module, it can answer frequently asked questions about drugs. In different modules, the content that can be replied is different. Thus, it is necessary to combine different response modules to deal with various questions.

## 3 Implementation

We implement each respond module in chatbot server, the operating system used to deploy chatbot is Ubuntu 16.04, and the development environment is python 3.6.

The BERT (Bidirectional Encoder Representations from Transformers) [20] model used in chatbot module is pre-trained by google-research (BERT-Base, Chinese: Chinese Simplified and Traditional, 12layer, 768hidden, 12heads, 110M parameters). BERT is a new method of pre-training language representations that has been trained a general-purpose "language understanding" model on a large text corpus, such as Wikipedia. Then we can use the pre-trained BERT model for downstream Natural Language Processing (NLP) tasks, the model can be fine-tuned with just one additional output layer to create state-of-the-art models for a wide range of tasks, such as question answering and language inference, without substantial task-specific architecture modifications. We utilized BERT to convert a user's query into a fixed-length sentence vector. In BERT, pooling is required to get a fixed representation of a sentence. Because a pre-trained model is not fine-tuned on any downstream tasks yet, in this case, the hidden state of [CLS] is not a good sentence representation. So we take the second-to-last hidden layer of all of the tokens in the sentence and do average pooling. In order to be able to connect with major social media platform, we need a web server. We have the website part of the Heroku cloud application platform and the development environment is GoLang.

**Fig. 6.** LineBot: Anti-Drug Buddy

Currently, Figure 6 presents our Anti-Drug Buddy is already serving on LINE, cooperate with addiction treatment doctors, and have more than 1100 users as friends. The current functions of our system have the following points: Information about drugs, How to deal with drug problems, Chitchat and Pushing the latest news and information about drugs. Figure 7 and 8 show our knowledge-based and retrieval-based modules dialogue. The knowledge-based can answer information about drugs. The retrieval-based can answer frequently asked questions about drugs in the domain of anti-drugs. In contrast, the domain is not in anti-drug that respond the chitchat corpus of PTT.

Moreover, we also provide additional anti-drug knowledge tests specially designed for users in elementary schools, junior high schools, and high schools (or

**Fig. 7.** Knowledge-based

**Fig. 8.** Retrieval-based

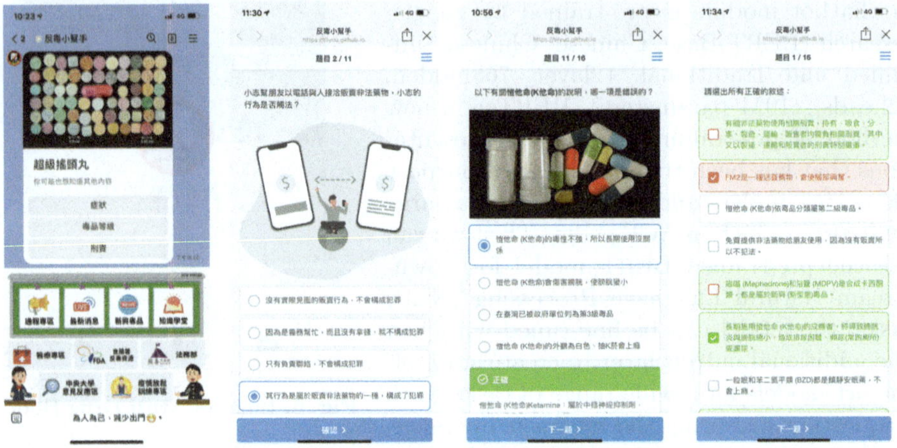

**Fig. 9.** Main view of Anti-Drug Buddy and anti-drug knowledge tests

above) with three different levels of difficulty. By recognizing and understanding these issues in the tests, people will be able to learn the correct knowledge about drugs and raise consciousness of anti-drug. Figure 9 displays main view of Anti-Drug Buddy and the example questions of anti-drug knowledge tests.

## 4 Conclusion and Future Works

In this paper, we propose Anti-Drug Buddy, which utilize chatbot to advocate general knowledge about drugs and deepening anti-drug awareness for people. Currently, our Anti-Drug Buddy is already serving on LINE. For an anti-drug domain, our chatbot can answer users: Information about drugs, How to deal with drug problems, Pushing the latest news and information about drugs. In order to expand and automatically generate knowledge graph, in the future, we will use semi-structured data such as the Wikipedia website to extract the structures as training corpus to train the machine learning model. Further, we will apply it to the entity attribute extraction of unstructured data.

On the other hand, we can analyze user queries. Our chatbot collects user conversations, which can be analyzed later. First, we can train a classification model to classify whether the user's query is an anti-drug domain. Then, try to find some features in these queries, get a tendency, or answer satisfaction.

## References

1. United Nations Office on Drugs and Crime: 2013 world drug report notes stability in use of traditional drugs and points to alarming rise in new psychoactive substances (2013). https://www.unodc.org/islamicrepublicofiran/en/2013-world-drug-report-notes-stability-in-use-of-traditional-drugs-and-points-to-alarming-rise-in-new-psychoactive-substances.html
2. Wakefield, M.A., Loken, B., Hornik, R.C.: Use of mass media campaigns to change health behaviour. Lancet **376**(9748), 1261–1271 (2010)
3. Huang, C.M., Wang, J.H., Chen, S.Y.: An investigation of a game-based anti-drug system: addictive learners vs. non-addictive learners. In: 2015 IEEE 15th International Conference on Advanced Learning Technologies, pp. 155–157. IEEE (2015)
4. Yang, T.C., Chen, M.C., Sun, Y.S.: An investigation of the influence of drug addiction on learning behaviors in a game-based learning environment. In: 2017 IEEE 17th International Conference on Advanced Learning Technologies (ICALT), pp. 158–162. IEEE (2017)
5. Ministry of Justice: The statistics on the number of drug offenders (2018). https://anti-drug.moj.gov.tw/Public/Charts/23/21a82272e583784130b63054594c5b0203.html
6. Shum, H.-Y., He, X., Li, D.: From eliza to xiaoice: challenges and opportunities with social chatbots, arXiv preprint arXiv:1801.01957 (2018)
7. Leviathan, Y., Matias, Y.: Engineering, Google, Google duplex: An ai system for accomplishing real-world tasks over the phone (2018). https://ai.googleblog.com/2018/05/duplex-ai-system-for-natural-conversation.html
8. Mnasri, M.: Recent advances in conversational nlp: Towards the standardization of chatbot building, arXiv preprint arXiv:1903.09025 (2019)
9. Gao, J., Galley, M., Li, L.: Neural approaches to conversational AI. In: The 41st International ACM SIGIR Conference on Research & Development in Information Retrieval, pp. 1371–1374 (2018)
10. Harms, J.-G., Kucherbaev, P., Bozzon, A., Houben, G.-J.: Approaches for dialog management in conversational agents. IEEE Internet Comput. **23**(2), 13–22 (2018)
11. Hussain, S., Athula, G.: Extending a conventional chatbot knowledge base to external knowledge source and introducing user based sessions for diabetes education. In: 2018 32nd International Conference on Advanced Information Networking and Applications Workshops (WAINA), pp. 698–703. IEEE (2018)
12. Athreya, R.G., Ngonga Ngomo, A.C., Usbeck, R.: Enhancing community interactions with data-driven chatbots-the dbpedia chatbot. In: Companion Proceedings of the The Web Conference 2018, pp. 143–146 (2018)
13. Kephart, J.O., Dibia, V.C., Ellis, J., Srivastava, B., Talamadupula, K., Dholakia, M.: An embodied cognitive assistant for visualizing and analyzing exoplanet data. IEEE Internet Comput. **23**(2), 31–39 (2019)
14. Cui, W., Xiao, Y., Wang, H., Song, Y., Hwang, S.-W., Wang, W.: Kbqa: learning question answering over qa corpora and knowledge bases, arXiv preprint arXiv:1903.02419 (2019)

15. Yu, Z., Papangelis, A., Rudnicky, A.: Ticktock: a non-goal-oriented multimodal dialog system with engagement awareness. In: 2015 AAAI Spring Symposium Series (2015)
16. Kannan, A., et al.: Smart reply: automated response suggestion for email. In: Proceedings of the 22nd ACM SIGKDD International Conference on Knowledge Discovery and Data Mining, pp. 955–964 (2016)
17. Sutskever, I., Vinyals, O., Le, Q.V.: Sequence to sequence learning with neural networks. In: Advances in Neural Information Processing Systems, pp. 3104–3112 (2014)
18. Hu, T., et al.: Touch your heart: a tone-aware chatbot for customer care on social media. In: Proceedings of the 2018 CHI Conference on Human Factors in Computing Systems, pp. 1–12 (2018)
19. Xu, B., et al.: CN-DBpedia: a never-ending Chinese knowledge extraction system. In: Benferhat, S., Tabia, K., Ali, M. (eds.) IEA/AIE 2017. LNCS (LNAI), vol. 10351, pp. 428–438. Springer, Cham (2017). https://doi.org/10.1007/978-3-319-60045-1_44
20. Devlin, J., Chang, M.W., Lee, K., Toutanova, K.: Bert: pre-training of deep bidirectional transformers for language understanding, arXiv preprint arXiv:1810.04805 (2018)

# Metagenome Analysis of Two River Microbial Flora for Modeling River Microbial Diversity

Jin Michel Ogawa and Ikuko Eguchi Yairi(✉)

Graduate School of Science and Engineering, Sophia University, 7-1 Kioicho, Chiyoda-ku, Tokyo 102-8554, Japan
ogawa@yairilab.net

**Abstract.** The microbiome drifting through the upper stream to downstream is considered to be affected by innumerable natural or artificial factors. Caching up the changing microbiome from these factors is essential for society to make the novel index of environment pollution and health safety. Since DNA sequencing technology improved these days, the number of studying river metagenome are increasing, and the feature of changing river microbiome are surveyed. However, the predictable models for river microbiome are still not understood yet. In this study, we found the influential factors of the microbiome from two rivers, Thames river, and Sinos river, examined the possibility of constructing predictable models for river microbiome by comparing with each river. Our result showed that both the Thames river and Sinos river indicate microbiome shift by longitudinal water flow. The result of the Thames river shows that the tributaries microbiome has a more negligible effect on the mainstream microbiome.

**Keywords:** Microbiome · Bioinformatics · Metagenome analysis · River

## 1 Introduction

Since the microbial flora of rivers is affected by countless potential events in the basin, such as domestic wastewater, precipitation, and inflow of surrounding soil, it has the potential to be an indicator of environmental pollution and the health of residents. In recent years, studies investigating the relationship between the microbial flora of rivers and the spatiotemporal characteristics around rivers have been conducted in various countries worldwide because identifying the factors that contribute to changes in the microbial flora has become an important social issue [1–6, 14, 15]. The succession of living organisms in a river is proposed in the "The river continuum concept" [7], and the structure and function of biological tissues in a river are determined in the flow of the river by chemical substances and the cycle of the river.

The survey in Thames river in England investigated how the microbial flora of bacterioplankton changes in a vast river flow [2]. They focused on the residence time of river water. They showed that the composition of bacterioplankton after merging with the tributary was different from that of the tributary and related to water's residence time. In a survey of the Danube River, which flows across Europe, they concluded that


Y. Takama et al. (Eds.): JSAI 2021, AISC 1423, pp. 213–223, 2022.
https://doi.org/10.1007/978-3-030-96451-1_19

the composition of Bacteria is due mainly to the influx of soil and groundwater ('mass effect) in the upstream, while that of mainly downstream depends on the significant bacteria constructed in the upper stream ('species sorting) [3]. Moreover, they considered decreasing alpha diversity of the microflora from upstream to downstream is due to the effects of these 'mass effect and 'species sorting. In the Sinos river, microbial flora in Brazil, the significant bacteria groups are maintained in the river flow and strongly support the stable Bacteria composition (seed-bank) in the source water [1].

From related studies, it is thought that the microbial flora transitions continuously from upstream to downstream. It might be possible to construct a river basin model for the microbial flora prediction. However, building a river basin model of this predictable microbial flora is an unknown region in river microbial flora studies. Therefore, in this study, we performed a metagenome analysis of the microbial flora of the Thames river and Sinos river, respectively, to investigate the factors that affect the microbial flora of the river.

## 2 Thames River Metagenome

### 2.1 Material and Method

The bacterial OTU data of the Thames river was taken from the paper titled "Catchment-scale biogeography of riverine bacterioplankton," which is posted in the academic journal ISME Journal ("Catchment-scale biogeography of riverine bacterioplankton Supplementary Table 2" is licensed by CC BY 3.0) [2]. The microbiome data was sampled on September 13th in 2011, at the west of London which is shown at Fig. 1. Samples were taken from 23 locations, including six main streams and 17 main tributaries. The number of reads was not nearby among sampling locations, so a rarefaction to normalize the

**Fig. 1.** The sampling points of Thames river. The map was created by "google mymap" referring to "Catchment-scale biogeography of riverine bacterioplankton Supplementary Table 1 [2]."

data was conducted. Name identifications at the Phylum level were conducted in each sample, and only bacteria were extracted from the OTU. For the fundamental analysis, a bar chart of the bacteria abundance ratio was created in each sample. OTU data that was not rarefied were used for the alpha diversity analysis, and Shannon's index, Pielou's evenness, Simpson's index, and Simpson's evenness were calculated [8, 9]. Analysis of beta diversity, Bray Curtis distance was calculated, and NMDS based on Bray Curtis distance was conducted to show the dissimilarity between samples [10–12].

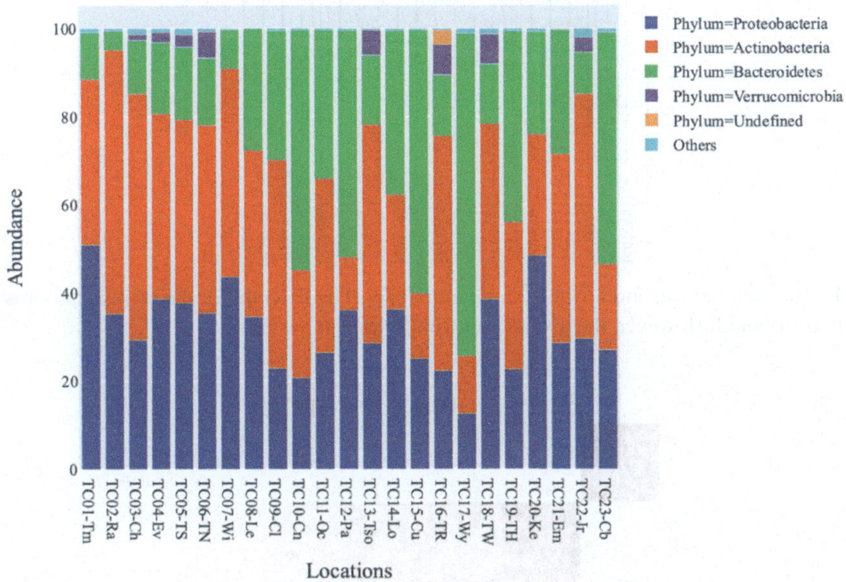


**Fig. 2.** The composition of bacteria. The main bacteria at the phylum level in the Thames river were Bacteroidetes, Proteobacteria, and Actinobacteria.

## 2.2 Result

The range of total reads was widely distributed from 2179 reads to 22907 reads. According to the bar chart in Fig. 2, the most three appeared bacteria in phylum level were Proteobacteria, Bacteroidetes, and Actinobacteria which occupied over 85% of all samples. Moreover, Verrucomicrobia has mainly emerged at mainstreams which are TC05-TS, TC06-TN, TC13-Tso, TC16-TR, TC18-TW. For alpha diversity indexes, shown in Fig. 3, there was substantially no difference between the mainstream locations except TC19-TH. For beta diversity, Bray Curtis distance, shown in Fig. 4, indicated that the similarity between mainstream locations was comparably high except for TC19-TH. NMDS, shown in Fig. 5, suggested that each bacteria composition of tributaries is unique, and bacteria in mainstreams are pretty similar to each other.

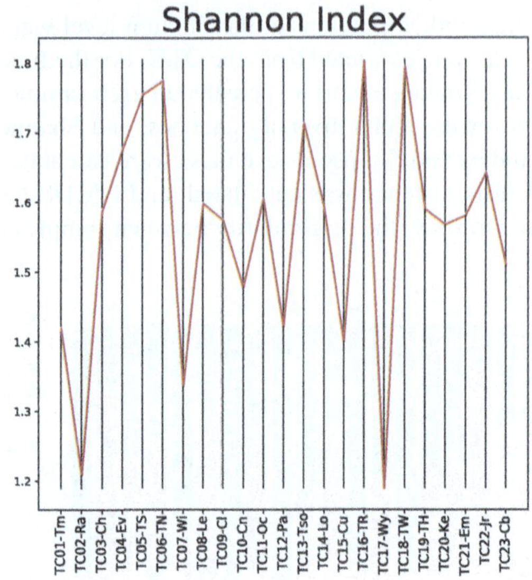


**Fig. 3.** The value of Shannon's index in phylum level. The diversity of the mainstream was not wholly distributed. However, the diversity of tributaries was various.

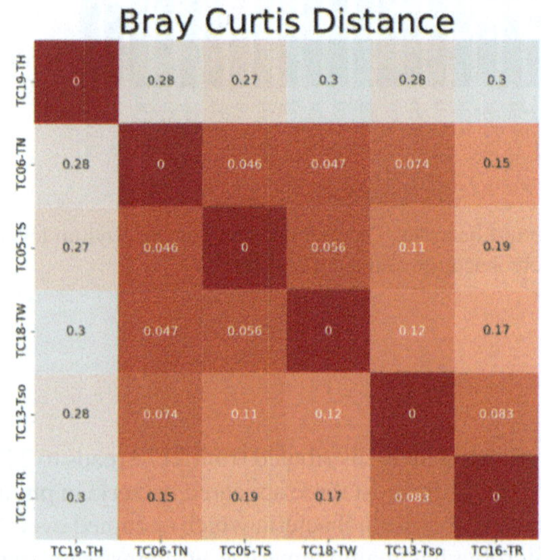


**Fig. 4.** Bray Curtis distance between each mainstream sampling point. Each of them had high similarity except for TC19-TH.

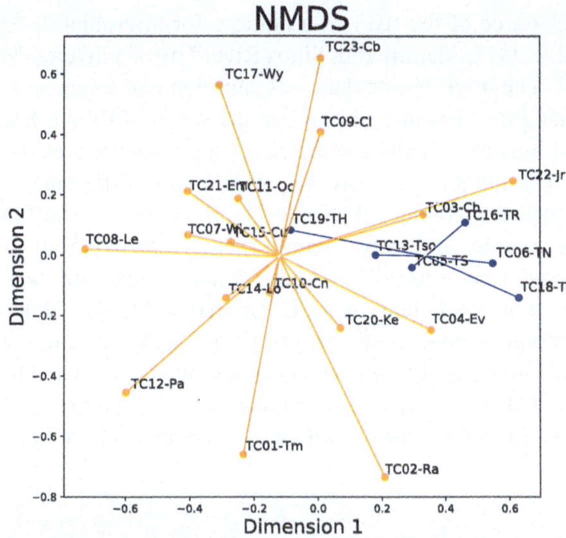


**Fig. 5.** NMDS connecting line by the sampling site in Thames river (stress = 0.1866). Mainstream sampling points were close together.

### 2.3 Discussion

Through the survey, the microbiomes of mainstreams are pretty similar. Indeed, the bacteria composition of all mainstreams has the same feature in the appearance of Virrucomicrobia and the low ratio of Bacteroidetes. This tendency indicates that the microbiome continually changes its composition. However, sampling point TC19-TH, where it seems the main channel, is quite different from other main streams. The main difference we find between TC19-TH and other mainstream sampling points is the amount of Verrucomicrobia. Verrucomicrobia was not found at TC19-TH, but it first appeared from TC09-Cl, a tributary of the Thames river upstream. Therefore, we consider that the tributaries microbiome would affect the mainstream microbiome.

Nevertheless, the result of NMDS shows that tributaries have various microbiomes, and most of their bacteria composition is utterly different from mainstreams. This indicates the possibility that the microbiome of tributaries located upstream affects the mainstream microbiome more than the downstream. Finally, it might be beneficial to find when the mainstream microbiome gets stable for investigating the factors which affect the bacterial composition. For that purpose, it is necessary to search both mainstream and tributary microbiome where it merges.

## 3 Sinos River Metagenome

### 3.1 Material and Method

The bacterial OTU data of Sinos river was taken from the paper titled "The Source of the River as a Nursery for Microbial Diversity," which is posted in the academic journal

PLOS ONE ("The source of the river as a nursery for microbial diversity. Supporting Information S2 Table. OTU identified in Sinos River" by de Oliveira, Margis is licensed by CC BY 4.0) [1]. The microbiome data was sampled two times on February 14th for summer and August 28th for winter in 2012 in the south of Brazil. Each time, samples were taken from 14 locations, including the clean upper stream area through to polluted downstream area. The number of reads was significantly different, so that rarefaction to normalize the data was conducted. Name identifications at the Phylum level were conducted in each sample, and only bacteria were extracted from the OTU. For the fundamental analysis, a bar chart of the bacteria abundance ratio was created in each sample, and correlation coefficients were calculated. OTU data that was not rarefied were used for the alpha diversity analysis, and Chao1 index, Shannon's index, Pielou's evenness, Simpson's index, and Simpson's evenness were calculated [8, 9]. Analysis of beta diversity, Bray Curtis distance was calculated, and NMDS based on Bray Curtis distance was conducted to show the dissimilarity between each sample [10–12] (Fig. 6).

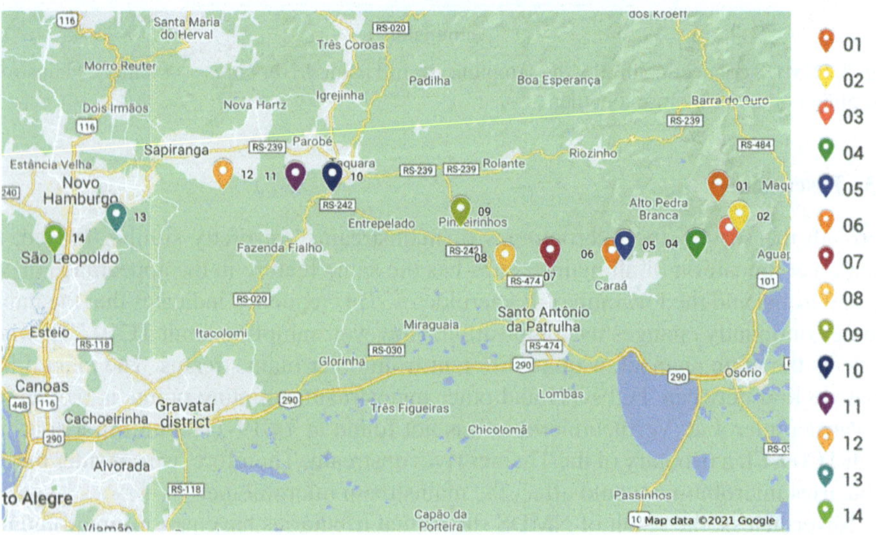


**Fig. 6.** The sampling points of Sinos river. The map was created by "google mymap", referring to "The source of the river as a nursery for microbial diversity. Figure 1" [1]

## 3.2 Result

The range of total reads was widely distributed from 3449 reads to 216211 reads. According to the bar chart in Fig. 7, the most two appeared bacteria in phylum level were Proteobacteria and Bacteroidetes, which occupied over 70% in all of the samples, and continue to Actinobacteria, Verrucomicrobia, and Cyanobacteria. The correlation heat map is shown in Fig. 8 indicated the high positive correlation between the samples, S07 to S09 and W07 to W09, but there was no correlation between the time. The number of unique bacteria and the value of the Choa1 index decreased along with the river flow through W01 to W08, while the survey in summer showed the opposite results. There was substantially no difference between the times in exact location for other alpha diversity indexes, only to find that the value of indexes is slightly higher in summer than in winter, from an example of Shannon's index shown in Fig. 9. For beta diversity, Bray Curtis distance indicated that the similarity between W01 to W03 and S05 to S07 was comparably high. NMDS, shown in Fig. 10, indicated that the bacteria composition between summer and winter was quite different. However, it showed a similarity in W01, W02, W03, S01, S02, and S04.

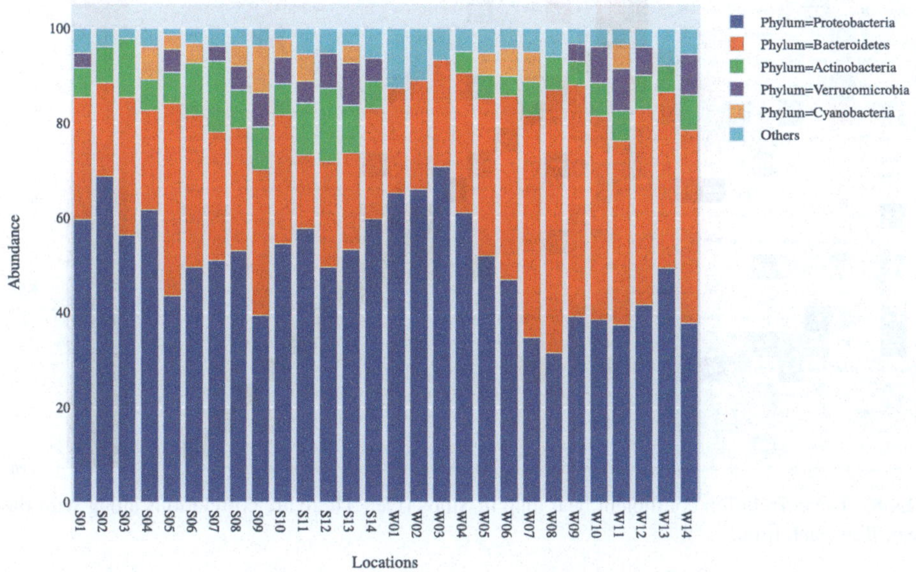


**Fig. 7.** The composition of bacteria. The main bacteria at the phylum level in the Sinos river were Bacteroidetes and Proteobacteria.

### 3.3 Discussion

The summer and winter bacterial composition in the Sinos river differed significantly from the correlation coefficient and NMDS results. PERMANOVA indicated a significant difference between the summer and winter bacteria composition at a significance level of 1% ($R2 = 0.199$; $p = 0.005$) [13]. However, from the results of Bray Curtis distance and NMDS, the similarity between 01 and 03 in the summer and winter upstream is high, and the number of species in phylum-level appearing in 01 is about the same in both summer and winter.

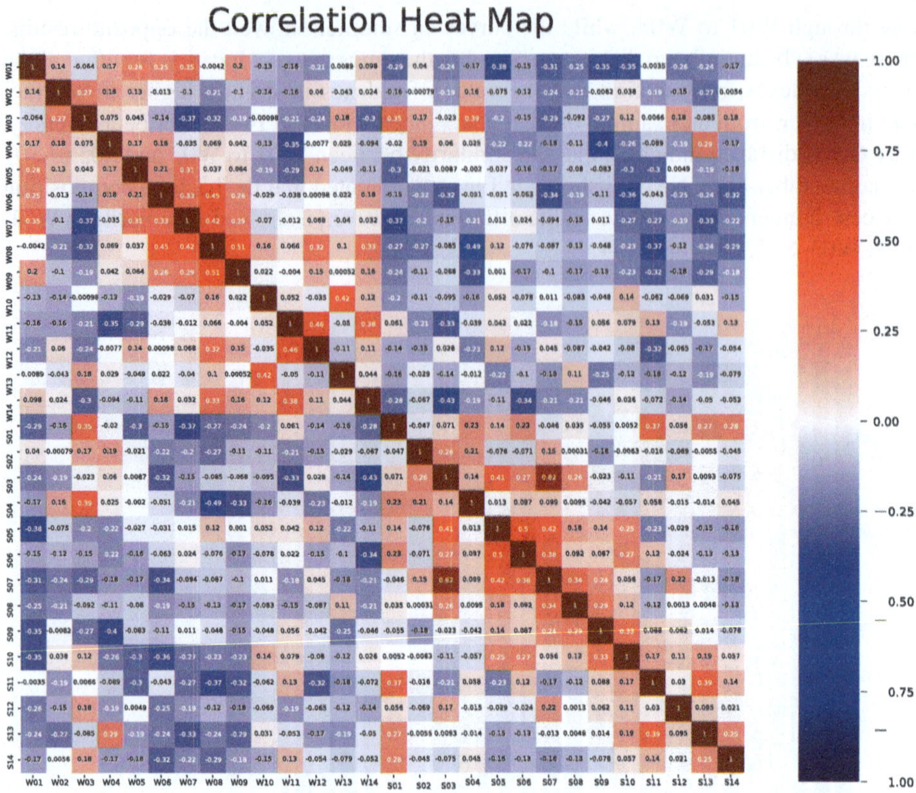


**Fig. 8.** The correlation coefficient heat map in Sinos river. There are correlations along with the river flow each time.

This indicates the possibility that the composition of bacteria is kept throughout the time in the headwaters of the Sinos river. This stable microbial flora in headwater was also found in the survey of the Mississippi river [5]. Therefore, the existence of normal microbial flora in the upper stream gets strong evidence, and these organizations might affect the microbiome downstream of each river.

## 4 Consideration for Model Development

In comparing the Thames river and the Sinos river, we performed a coverage-based rarefaction to modify both samples [14]. For comparison, Bray Curtis distance-based NMDS were implemented. From NMDS in Fig. 11, it indicated the bacterial composition of the Thames river and Sinos river. Moreover, the result shows tributary points had various bacterial compositions from it widely distributed in the graph. However, it is interesting that even tributaries bacterial composition is diverse, our survey shows the continual transition of the microbiome through the mainstream in both Thames river and Sinos River. This fact indicates that, at a point in the river, the microbiome is not affected by the tributaries, but the main course of the upper stream affects its microbiome. We

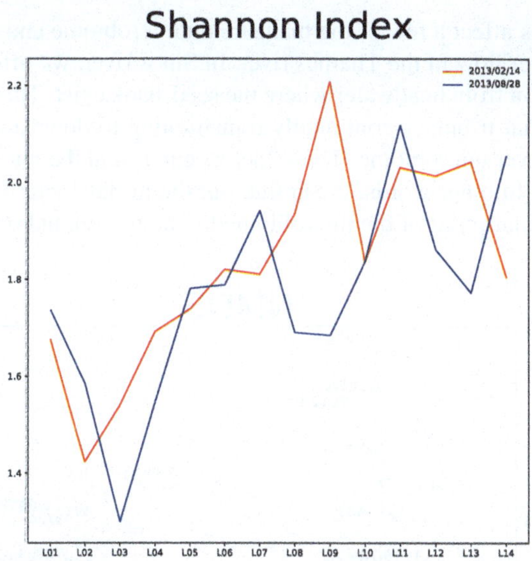


**Fig. 9.** The value of Shannon's index in rarefied bacteria in phylum level. Summer has a higher value than winter.

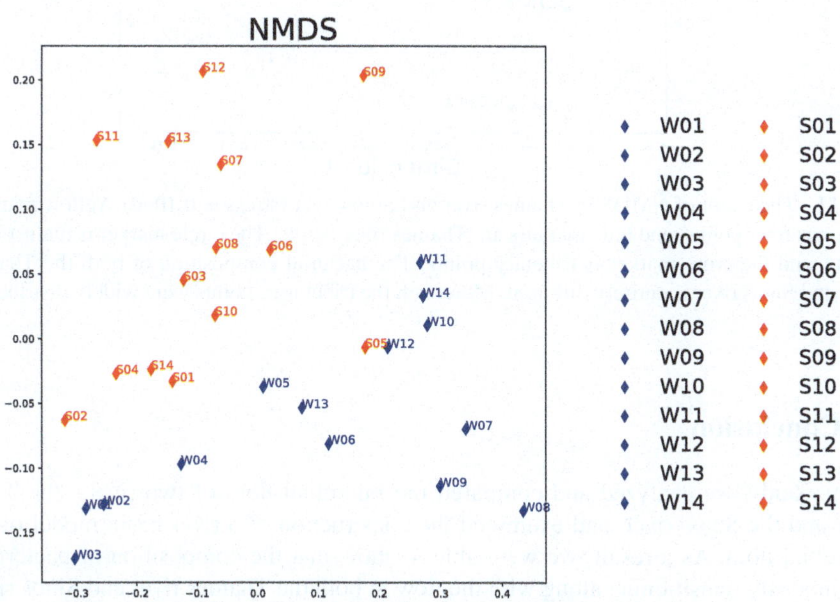


**Fig. 10.** The result of NMDS in the Sinos river (stress = 0.066). The separation between summer and winter data indicates that bacteria composition continuously changes each time rather than randomly.

show that tributaries affect a few of the mainstream microbiome and continually transition through the river flow in the Thames river. In Sinos river, we offer that the bacteria continually transition from headwater where the seed-bank exist. These facts would link to the hypothesis that tributary continually transitioning to downstream factors would involve modeling river microbiome. These factors can reveal the main stable headwater of points in a river. In other words, it can find out the initial bacterial composition and calculate the particular point of the microbiome by their continual transition feature.

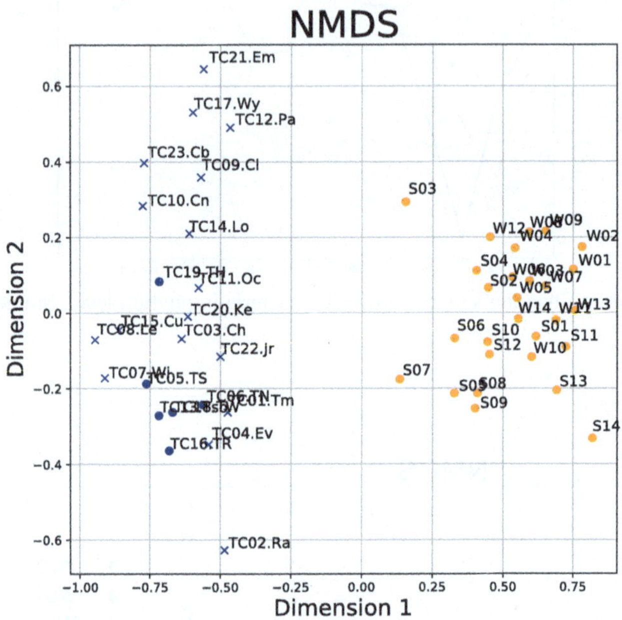


**Fig. 11.** The result of NMDS in Thames river and Sinos river (stress = 0.1060). Yellow markers are Sinos river points, and blue markers are Thames river points. The circle marker is mainstream points, and the cross marker is tributary points. The bacterial composition of both the Thames river and Sinos river is entirely different. Moreover, the tributaries points were widely distributed.

## 5 Conclusion

In this study, we analyzed and compared the microbial flora of two rivers, the Tama river and the Sinos river, and examined the construction of a river basin model of the microbial flora. As a result, we were able to show that the composition of bacteria is continuously transitioning along with the flow in both the Thames river and Sinos river. In the Sinos river, there is a seed bank which stable form the microbiome. This result agrees with "The river continuum concept" [7]. In the future, we plan to build a model by mentioning mering points between mainstream and tributaries and finding factors that maintain the stability of the bacterial composition which flows downstream.

**Acknowledgements.** We are grateful to all of Yairi laboratory bio-team members for assisting our research.

## References


1. De Oliveira, L.F., Margis, R.: The source of the river as a nursery for microbial diversity. PLOS ONE (2015)
2. Read, D.S., et al.: Catchment-scale biogeography of riverine bacterioplankton. ISME J. **9** (2015)
3. Savio, D., et al.: Bacterial diversity along a 2600 km river continuum. Environ. Microbiol. (2015)
4. Zancarini, A., Echenique-Subiabre, I., Debroas, D., et al.: Deciphering biodiversity and interactions between bacteria and microeukaryotes within epilithic biofilms from the Loue River, France. Sci. Rep. **7** (2017)
5. Staley, C., et al.: Application of illumina next-generation sequencing to characterize the bacterial community of the Upper Mississippi River. J. Appl. Microbiol. (2013)
6. Zeglin, L.: Stream microbial diversity in response to environmental changes: review and synthesis of existing research. Front. Microbiol. (2015)
7. Vannote, R.L., Minshall, G.W., Cummins, K.W., Sedell, J.R., Cushing, C.E.: The river continuum concept. Can. J. Fish. Aquat. Sci. (1980)
8. Chao, A.: Non-parametric estimation of the classes in a population. Scand. J. Stat. (1984)
9. Lande, R.: Statistics and partitioning of species diversity and similarity among multiple communities. Oikos (1996)
10. Oksanen, J., et al.: Package vegan. Community ecology package, version 2.9 (2013)
11. Clarke, K.R.: Non-parametric multivariate analyses of changes in community structure. Australian J. Ecol. **18**(1) (1993)
12. Manly, B.F., Alberto, J.A.N.: Multivariate Statistical Methods: A Primer. CRC press, Boca Raton (2016)
13. Anderson, M.: A new method for non-parametric multivariate analysis of variance. Austral Ecol. (2001)
14. Chao, A., Jost, L.: Coverage-based rarefaction and extrapolation: standardizing samples by completeness rather than size. Ecology **93**, 2533–47 (2012). https://doi.org/10.2307/41739612
15. Ghai, R., Rodriguez-Valera, F., McMahon, K.D., et al.: Metagenomics of the water column in the pristine upper course of the Amazon river. PLoS One (2011)

# Author Index





Y. Takama et al. (Eds.): JSAI 2021, AISC 1423, pp. 225–226, 2022.
https://doi.org/10.1007/978-3-030-96451-1

The manufacturer's authorised representative in the EU is Springer Nature Customer Service Centre GmbH, Europaplatz 3, 69115 Heidelberg, Germany. If you have any concerns regarding our products, please contact ProductSafety@springernature.com

Printed and bound by CPI Group (UK) Ltd, Croydon, CR0 4YY
30/09/2026
02225320-0005